Introduction to Matrix Theory and Linear Algebra

Irving Reiner
University of Illinois

HOLT, RINEHART AND WINSTON, INC.
New York Chicago San Francisco Atlanta
Dallas Montreal Toronto London Sydney

To Irma

Library of Congress Catalog Card Number: 79-151082
AMS 1970 Subject Classification: 1501
SBN: 03-085410-5
Printed in the United States of America
1234 090 987654321

Preface

This book gives a quick introduction to the basic ideas and calculations of matrix theory. It may be used as supplementary material in the second or third semester of a standard calculus course, and can be covered in 15 to 20 class hours. Optional topics and optional proofs may be omitted if desired without affecting the treatment of most of the material following such topics. **Optional material will be enclosed in brackets *[. . .]* throughout the book.**

Rather than stressing proofs of theorems, I have tried to emphasize basic concepts and manipulative skills. Proofs are often given in special cases, working sometimes in two- or three-dimensional space for simplicity. For more detailed accounts of matrix theory and linear algebra, the reader may consult *Linear Algebra* by C. W. Curtis, *Elementary Matrix Algebra* by F. E. Hohn, or *Theory of Matrices* by S. Perlis. These are three of many texts on these subjects.

The student should be slightly familiar with determinants, although all of the necessary results on determinants are reviewed in Section 5. Hopefully the student knows what is meant by the sum of two vectors and by a scalar multiple of a vector. These ideas are convenient for motivating certain matrix definitions and are reviewed briefly when first introduced. Some results from calculus are needed in Section 8, and partial derivatives occur in Section 15. It would be possible to use this book in a precalculus course by omitting part of Section 8 and all of Section 15.

The numbers and scalar quantities which occur in the text are usually assumed to be real numbers, although almost all of the results obtained are equally valid when complex numbers are used. In Sections 11 to 15, the student's attention is restricted to vectors and matrices whose entries are real numbers. Such additional restrictions are always stated explicitly when they are needed.

Starred problems are more difficult than average. Solutions of some of the exercises are given at the end of the book.

Irving Reiner

Urbana, Illinois
March, 1971

MRS. Mcpherson

Wal. 201

Thur 4:00PM

Contents

1 INTRODUCTION

A *matrix* is a set of numbers (real or complex) arranged in a rectangular array. For example, consider the matrix

$$A = \begin{bmatrix} 2 & 4 & -1 \\ 3 & 2 & 0 \end{bmatrix}. \tag{1.1}$$

It has two rows: the first row consists of the numbers $2, 4, -1$; the second row is $3, 2, 0$. The matrix has three columns: the first column is $2, 3$, the second column is $4, 2$, and the third column is $-1, 0$. We call A a *two by three* matrix, and write $A^{2\times 3}$ when we wish to emphasize the fact that A has 2 rows and 3 columns.

Other examples of matrices are

$$\begin{bmatrix} 1 & 1 & 1 \\ 1 & 1 & 1 \\ 1 & 1 & 1 \end{bmatrix}, \quad [6], \quad \begin{bmatrix} 5 & 1 \\ 4 & 2 \\ 3 & 3 \\ 2 & 4 \end{bmatrix}, \quad \begin{bmatrix} a & b & c \\ d & e & f \end{bmatrix}, \quad [x \quad y \quad z].$$

These are of sizes 3×3, 1×1, 4×2, 2×3, and 1×3, respectively. In general, an $m \times n$ matrix has m rows and n columns. A *square* matrix is a matrix with the same number of rows and columns. Square matrices can be of sizes 1×1, 2×2, 3×3, and so on.

Caution: A square matrix is *not* the same thing as a determinant. Recall that a determinant is a *number* gotten by performing certain operations on a square array. For example, if

$$A = \begin{bmatrix} a & b \\ c & d \end{bmatrix},$$

then A is a 2×2 matrix, and the number $ad - bc$ is called the determinant of A. In Section 5 we shall review some properties of determinants.

In the matrix A defined in (1.1), the number 4 occurs in the first row and second column. We call 4 the (1,2)-entry of A, and we say that 4 is in the (1,2)-position of A. Similarly, -1 is the (1,3)-entry of A, and 0 is the (2,3)-entry of A.

In the matrix

$$\begin{bmatrix} 5 & 1 \\ 4 & 2 \\ 3 & 3 \\ 2 & 4 \end{bmatrix},$$

the (1,1)-entry is 5, the (2,1)-entry is 4, the (4,1)-entry is 2, the (1,2)-entry is 1, the (3,2)-entry is 3, and so on.

To write down a 3×3 matrix with variable entries, it is often convenient to use double subscripts. For example, we can denote the (1,1)-entry by the symbol a_{11}, the (2,1)-entry by a_{21}, and so on. Thus a general 3×3 matrix can be written as

$$A = \begin{bmatrix} a_{11} & a_{12} & a_{13} \\ a_{21} & a_{22} & a_{23} \\ a_{31} & a_{32} & a_{33} \end{bmatrix}.$$

The first subscript of each entry tells us in which *row* the element occurs, and the second subscript tells us in which *column* the element lies. Thus a_{32} occurs in the third row and second column of A; that is, a_{32} is the (3,2)-entry of a.

A general 2×4 matrix can be written down as

$$\begin{bmatrix} a_{11} & a_{12} & a_{13} & a_{14} \\ a_{21} & a_{22} & a_{23} & a_{24} \end{bmatrix}.$$

Where do matrices come from? Often they arise from systems of simultaneous linear equations. For example, the system

(1.2)
$$\begin{cases} 2x + 3y + 4z = 7 \\ x - \ \ y + 3z = 4 \end{cases}$$

gives rise to a *matrix of coefficients* (of the unknowns x, y, z), namely

$$\begin{bmatrix} 2 & 3 & 4 \\ 1 & -1 & 3 \end{bmatrix}.$$

One may also form an *augmented* matrix by adjoining an extra column which consists of the constant terms on the right in (1.2). In this case, the augmented matrix is

$$\begin{bmatrix} 2 & 3 & 4 & 7 \\ 1 & -1 & 3 & 4 \end{bmatrix}.$$

As we shall see later on, matrices are a convenient device for handling systems of linear equations.

2 OTHER NOTATIONS FOR MATRICES

We have described a matrix as a rectangular array of numbers, enclosed in brackets for convenience. Some authors prefer to use parentheses in place of brackets, and write a 2×3 matrix (for example) as

$$A = \begin{pmatrix} a_{11} & a_{12} & a_{13} \\ a_{21} & a_{22} & a_{23} \end{pmatrix}.$$

In older books, one also finds the notation

$$A = \begin{Vmatrix} a_{11} & a_{12} & a_{13} \\ a_{21} & a_{22} & a_{23} \end{Vmatrix}.$$

In order to write down a general 4×5 matrix, we would need to write down 20 terms; the first row might be denoted by $a_{11}, a_{12}, a_{13}, a_{14}, a_{15}$; the second row by $a_{21}, a_{22}, a_{23}, a_{24}, a_{25}$; and so on. One possible abbreviation would be to write the matrix as

$$A = \begin{bmatrix} a_{11} & \cdots & a_{15} \\ \cdot & \cdots & \cdot \\ a_{41} & \cdots & a_{45} \end{bmatrix},$$

indicating only the corner entries. An even simpler notation would be as follows:

$$A = [a_{ij}]^{4\times 5}.$$

This means that the row subscript i ranges from 1 to 4, and the column subscript j ranges from 1 to 5.

As examples, note that

$$[a_{ij}]^{1\times 1} = [a_{11}], \qquad [b_{ij}]^{2\times 4} = \begin{bmatrix} b_{11} & b_{12} & b_{13} & b_{14} \\ b_{21} & b_{22} & b_{23} & b_{24} \end{bmatrix},$$

$$[c_{ij}]^{2\times 2} = \begin{bmatrix} c_{11} & c_{12} \\ c_{21} & c_{22} \end{bmatrix}.$$

EXERCISES

1. Give some examples of 1×2, 3×2, 4×2, 7×1, 1×5 matrices. Also give an example of an array of numbers which is *not* a matrix.
2. Write down a general 2×5 matrix.
3. Write out the matrices indicated by the following abbreviations:

$$[a_{ij}]^{1\times 4},\ [b_{ij}]^{2\times 3},\ [c_{ij}]^{3\times 1}.$$

4. Write out the matrix $[a_{ij}]^{3\times 4}$ in which $a_{ij} = 2i + j - 1$.
5. Write out the matrix $[\delta_{ij}]^{4\times 5}$, where

$$\delta_{ij} = \begin{cases} 1 \text{ when } j = i \\ 0 \text{ when } j \neq i \end{cases} \qquad (1 \leqslant i \leqslant 4).$$

3 OPERATIONS WITH MATRICES (SUM, SCALAR MULTIPLE, PRODUCT, TRANSPOSE)

(i) We shall say that two matrices are *equal* if and only if they are identical. Thus, $A = B$ means that the matrices A and B are of the same size, and that each entry of A equals the corresponding entry of B.

(ii) Let us start with a 1×3 matrix $[a \quad b \quad c]$, often called a *row vector.*[1] As usual, we may represent this vector geometrically in 3-dimensional space, by drawing a vector $\overrightarrow{OP}$ which starts at the origin O, and ends at the point P whose coordinates are (a,b,c). (See Figure 3.1.) If $[a' \quad b' \quad c']$ is another row vector, it can be represented geometrically by the vector $\overrightarrow{PQ}$ starting at P, and ending at the point Q whose coordinates are $(a + a',$ $b + b', c + c')$. The vector $\overrightarrow{OQ}$ is defined to be the *sum* of the two row

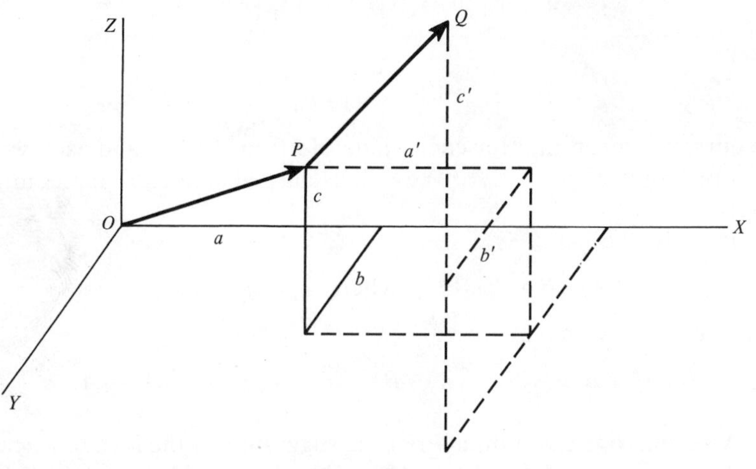

Figure 3.1

[1] The concept of a vector is used only to motivate the later definition of the sum of two matrices. The reader who is not familiar with this concept will find that the discussion is self-explanatory.

vectors $[a \quad b \quad c]$ and $[a' \quad b' \quad c']$ and we shall write

$$[a \quad b \quad c] + [a' \quad b' \quad c'] = [a + a' \quad b + b' \quad c + c'].$$

This suggests the following definition: if A and B are a pair of $m \times n$ matrices, their *sum* $A + B$ is the $m \times n$ matrix each of whose entries is the sum of the corresponding entries of A and B. For example,

$$A = \begin{bmatrix} a_{11} & a_{12} & a_{13} \\ a_{21} & a_{22} & a_{23} \end{bmatrix}, \qquad B = \begin{bmatrix} b_{11} & b_{12} & b_{13} \\ b_{21} & b_{22} & b_{23} \end{bmatrix},$$

$$A + B = \begin{bmatrix} a_{11} + b_{11} & a_{12} + b_{12} & a_{13} + b_{13} \\ a_{21} + b_{21} & a_{22} + b_{22} & a_{23} + b_{23} \end{bmatrix}.$$

In other words, if a_{ij} is the (i,j)-entry of A, and b_{ij} is the (i,j)-entry of B, then $a_{ij} + b_{ij}$ is the (i,j)-entry of $A + B$.

Caution: You can't add two matrices of different sizes.

EXAMPLES

$$\begin{bmatrix} 2 & 4 \\ -1 & 6 \end{bmatrix} + \begin{bmatrix} 3 & 2 \\ 1 & 1 \end{bmatrix} = \begin{bmatrix} 5 & 6 \\ 0 & 7 \end{bmatrix}, \qquad \begin{bmatrix} 4 \\ -3 \end{bmatrix} + \begin{bmatrix} -1 \\ 7 \end{bmatrix} = \begin{bmatrix} 3 \\ 4 \end{bmatrix},$$

$$[5] + [4] = [9], \qquad [2 \quad 1] + [3 \quad 1] = [5 \quad 2].$$

We can rephrase the definition of the sum of two matrices as follows: given matrices

$$A = [a_{ij}]^{m \times n}, \qquad B = [b_{ij}]^{m \times n},$$

we set

$$A + B = [c_{ij}]^{m \times n}, \text{ where } c_{ij} = a_{ij} + b_{ij}.$$

(Of course we mean that for each value of i from 1 to m, and each value of j from 1 to n, $c_{ij} = a_{ij} + b_{ij}$. We will usually omit such remarks in the future.)

Similarly, we put

$$A - B = [d_{ij}]^{m \times n}, \text{ where } d_{ij} = a_{ij} - b_{ij}.$$

Then in general we have

$$A + B = B + A, \qquad (A - B) + B = B + (A - B) = A.$$

(iii) A second operation on matrices is suggested by the idea of a scalar multiple of a vector, defined thus: let α be a scalar (that is, a number), and let $[a \quad b \quad c]$ be a row vector. We define

$$\alpha[a \quad b \quad c] = [\alpha a \quad \alpha b \quad \alpha c].$$

Let us represent $[a \quad b \quad c]$ by $\overrightarrow{OP}$, as before, and let P' be the point with coordinates $(\alpha a, \alpha b, \alpha c)$. If $\alpha > 0$, then $\overrightarrow{OP'}$ is gotten from $\overrightarrow{OP}$ by magnifying it by a factor of α, and we shall write $\overrightarrow{OP'} = \alpha\, \overrightarrow{OP}$. If $\alpha < 0$, then $\overrightarrow{OP'}$ has length $|\alpha|$ times the length of $\overrightarrow{OP}$, but is opposite to $\overrightarrow{OP}$ in direction.

This leads us to the following definition: let A be an $m \times n$ matrix, and let α be a number. Then αA is the *scalar multiple* of A gotten by multiplying each entry of A by α. Thus

$$A = [a_{ij}]^{m \times n}, \qquad \alpha A = [\alpha a_{ij}]^{m \times n}.$$

EXAMPLES

$$A = \begin{bmatrix} 2 & 4 \\ 1 & 3 \end{bmatrix}, \quad 2A = \begin{bmatrix} 4 & 8 \\ 2 & 6 \end{bmatrix}, \quad -A = \begin{bmatrix} -2 & -4 \\ -1 & -3 \end{bmatrix}, \quad -3A = \begin{bmatrix} -6 & -12 \\ -3 & -9 \end{bmatrix}.$$

$$4[1 \quad 2 \quad -1 \quad 7] = [4 \quad 8 \quad -4 \quad 28].$$

$$\left(\frac{-5}{2}\right)\begin{bmatrix} 4 \\ 2 \\ 0 \end{bmatrix} = \begin{bmatrix} -10 \\ -5 \\ 0 \end{bmatrix}.$$

There are obvious formulas, such as

$$A + A = 2A, \qquad A + 2A = 3A, \qquad A - 2B = A + (-2)B, \qquad -(-A) = A.$$

In short, matrices behave very much like ordinary numbers in all formulas involving addition, subtraction, and scalar multiples.

(iv) A more complicated operation is that of multiplication of matrices. How shall we define the *product* AB of two matrices A and B? The answer is suggested by an example involving linear equations. Consider the problem of solving the simultaneous linear equations

$$\begin{cases} 2x - 3y = 7 \\ \;\; x + 6y = 9 \end{cases}$$

for the unknowns x, y. The matrix of coefficients of the unknowns is defined to be

$$A = \begin{bmatrix} 2 & -3 \\ 1 & 6 \end{bmatrix}.$$

We wish to apply this matrix to the pair x, y so as to get the new pair $2x - 3y$, $x + 6y$. Let us write the pair x, y as a column vector $\begin{bmatrix} x \\ y \end{bmatrix}$; then we would like to obtain a new column vector $\begin{bmatrix} 2x - 3y \\ x + 6y \end{bmatrix}$ by somehow applying the matrix A to the original column vector $\begin{bmatrix} x \\ y \end{bmatrix}$. We shall write

$$\begin{bmatrix} 2 & -3 \\ 1 & 6 \end{bmatrix} \begin{bmatrix} x \\ y \end{bmatrix} = \begin{bmatrix} 2x - 3y \\ x + 6y \end{bmatrix}.$$

This suggests the following definition for the product of a 2×2 matrix and a column vector:

$$\begin{bmatrix} a_1 & a_2 \\ b_1 & b_2 \end{bmatrix} \begin{bmatrix} x \\ y \end{bmatrix} = \begin{bmatrix} a_1x + a_2y \\ b_1x + b_2y \end{bmatrix}.$$

The product is again a column vector.

By analogy, in three dimensions we put

$$\begin{bmatrix} a_1 & a_2 & a_3 \\ b_1 & b_2 & b_3 \\ c_1 & c_2 & c_3 \end{bmatrix} \begin{bmatrix} x \\ y \\ z \end{bmatrix} = \begin{bmatrix} a_1x + a_2y + a_3z \\ b_1x + b_2y + b_3z \\ c_1x + c_2y + c_3z \end{bmatrix}.$$

Thus, a 3×3 matrix times a 3×1 matrix gives a 3×1 matrix. However, we may want to apply these procedures not only to one column vector, but say to a pair of column vectors. We shall write

$$\begin{bmatrix} a_1 & a_2 & a_3 \\ b_1 & b_2 & b_3 \\ c_1 & c_2 & c_3 \end{bmatrix} \begin{bmatrix} x & x' \\ y & y' \\ z & z' \end{bmatrix} = \begin{bmatrix} a_1x + a_2y + a_3z & a_1x' + a_2y' + a_3z' \\ b_1x + b_2y + b_3z & b_1x' + b_2y' + b_3z' \\ c_1x + c_2y + c_3z & c_1x' + c_2y' + c_3z' \end{bmatrix}.$$

In this case, a 3×3 matrix times a 3×2 matrix gives a 3×2 matrix. Note that the (1,1)-entry of the product uses the first row of the 3×3 matrix, and the first column of the 3×2 matrix. Likewise, the (2,1)-entry of the product uses the second row of the 3×3 matrix, and the first column of the 3×2 matrix.

We are now ready to give the general definition. Let A be an $m \times n$ matrix, and let B be an $n \times p$ matrix. Then the *product* AB is an $m \times p$ matrix; the (i,j)-entry of AB uses the ith row of A and the jth column of B, and is defined by

$$(i,j)\text{-entry of } AB = a_{i1}b_{1j} + a_{i2}b_{2j} + \cdots + a_{in}b_{nj}.$$

The following will illustrate:

$$\begin{array}{c} \\ \\ i\text{th row} \rightarrow \\ \\ \\ \end{array} \begin{bmatrix} a_{11} & \cdots & a_{1n} \\ \cdot & \cdots & \cdot \\ a_{i1} & \cdots & a_{in} \\ \cdot & \cdots & \cdot \\ a_{m1} & \cdots & a_{mn} \end{bmatrix} \begin{bmatrix} b_{11} & \cdots & b_{1j} & \cdots & b_{1p} \\ \cdot & \cdots & \cdot & \cdots & \cdot \\ b_{n1} & \cdots & b_{nj} & \cdots & b_{np} \end{bmatrix}.$$

(jth column indicated by arrow above b_{1j}.)

Let us rephrase the definition: let

$$A = [a_{ij}]^{m\times n}, \qquad B = [b_{ij}]^{n\times p}.$$

Then

$$AB = [c_{ij}]^{m\times p}, \text{ where } c_{ij} = a_{i1}b_{1j} + a_{i2}b_{2j} + \cdot\cdot\cdot + a_{in}b_{nj}.$$

EXAMPLES

$$\begin{bmatrix} a & b \\ c & d \end{bmatrix}\begin{bmatrix} a' & b' \\ c' & d' \end{bmatrix} = \begin{bmatrix} aa' + bc' & ab' + bd' \\ ca' + dc' & cb' + dd' \end{bmatrix}.$$

$$[2 \ \ 3 \ \ 1]\begin{bmatrix} 4 & 3 \\ 1 & -1 \\ 6 & 5 \end{bmatrix} = [2\cdot 4 + 3\cdot 1 + 1\cdot 6, 2\cdot 3 + 3(-1) + 1\cdot 5] = [17 \ \ 8].$$

$$\begin{bmatrix} a & b \\ c & d \end{bmatrix}\begin{bmatrix} x \\ y \end{bmatrix} = \begin{bmatrix} ax + by \\ cx + dy \end{bmatrix}; \qquad [x \ \ y]\begin{bmatrix} a & b \\ c & d \end{bmatrix} = [xa + yc, xb + yd].$$

$$\begin{bmatrix} a & 0 & 0 \\ 0 & b & 0 \\ 0 & 0 & c \end{bmatrix}\begin{bmatrix} x & x' \\ y & y' \\ z & z' \end{bmatrix} = \begin{bmatrix} ax & ax' \\ by & by' \\ cz & cz' \end{bmatrix}.$$

$$\begin{bmatrix} 2 & 1 & 3 \\ 0 & 2 & 5 \end{bmatrix}\begin{bmatrix} 1 & 2 & 3 \\ 6 & -1 & 2 \\ 9 & 4 & 1 \end{bmatrix}$$

$$= \begin{bmatrix} 2\cdot 1 + 1\cdot 6 + 3\cdot 9, & 2\cdot 2 + 1(-1) + 3\cdot 4, & 2\cdot 3 + 1\cdot 2 + 3\cdot 1 \\ 0\cdot 1 + 2\cdot 6 + 5\cdot 9, & 0\cdot 2 + 2(-1) + 5\cdot 4, & 0\cdot 3 + 2\cdot 2 + 5\cdot 1 \end{bmatrix}$$

$$= \begin{bmatrix} 35 & 15 & 11 \\ 57 & 18 & 9 \end{bmatrix}.$$

Caution: The product AB is defined only when the number of columns in A equals the number of rows in B.

EXAMPLE

Let $A = [a_{ij}]^{5\times 3}$, and $B = [b_{ij}]^{3\times 9}$. Then the (2,5)-entry of AB is

$$a_{21}b_{15} + a_{22}b_{25} + a_{23}b_{35}.$$

The (1,7)-entry of AB is

$$a_{11}b_{17} + a_{12}b_{27} + a_{13}b_{37}.$$

The (i,j)-entry of AB is

$$a_{i1}b_{1j} + a_{i2}b_{2j} + a_{i3}b_{3j}.$$

Note that BA is not even defined. It is also easy to give examples where both AB and BA are defined, but are not equal to one another.

(v) We conclude with one last operation on matrices. If A is an $m \times n$ matrix, we may form a new matrix of size $n \times m$ called the *transpose* of A, by interchanging the rows and columns of A. Denote the transpose of A by A^{T} (or sometimes by A'). Then the first column of A^{T} is the first row of A, the second column of A^{T} is the second row of A, and so on.

EXAMPLES

$$A = \begin{bmatrix} a & b \\ c & d \end{bmatrix}, \qquad A^{\mathrm{T}} = \begin{bmatrix} a & c \\ b & d \end{bmatrix}.$$

$$B = \begin{bmatrix} a & b & c \\ d & e & f \end{bmatrix}, \qquad B^{\mathrm{T}} = \begin{bmatrix} a & d \\ b & e \\ c & f \end{bmatrix}.$$

$$C = [a \quad b \quad c], \qquad C^{\mathrm{T}} = \begin{bmatrix} a \\ b \\ c \end{bmatrix}.$$

$$D = \begin{bmatrix} c_{11} & c_{12} & \cdots & c_{15} \\ c_{21} & c_{22} & \cdots & c_{25} \\ \cdot & \cdot & \cdots & \cdot \\ c_{61} & c_{62} & \cdots & c_{65} \end{bmatrix}^{6\times 5}, \qquad D^{\mathrm{T}} = \begin{bmatrix} c_{11} & c_{21} & \cdots & c_{61} \\ c_{12} & c_{22} & \cdots & c_{62} \\ \cdot & \cdot & \cdots & \cdot \\ c_{15} & c_{25} & \cdots & c_{65} \end{bmatrix}^{5\times 6}.$$

In general, if $A = [a_{ij}]^{m\times n}$, then A^{T} is the $n \times m$ matrix whose (i,j)-entry is a_{ji}.

EXERCISES

1. Given the matrices

$$A = \begin{bmatrix} 3 & 2 \\ 4 & -1 \\ 6 & 1 \end{bmatrix}, \quad B = \begin{bmatrix} 2 & 5 \\ -1 & 4 \\ 0 & 3 \end{bmatrix}, \quad C = \begin{bmatrix} 5 & 7 \\ 1 & 2 \end{bmatrix}, \quad D = \begin{bmatrix} 2 & 1 & 3 \\ -1 & 4 & 1 \end{bmatrix},$$

$$E = [2], \quad F = [2 \quad -1], \quad G = \begin{bmatrix} 3 \\ 1 \end{bmatrix}, \quad H = \begin{bmatrix} 1 & 1 & 0 \\ 0 & 1 & 0 \\ 0 & 0 & 1 \end{bmatrix}.$$

Which sums of matrices are defined? Which products?

Calculate each of the following:

$$A + B,\ A + 2B,\ B - A,\ AC,\ AD,\ EF,\ FG,\ DH,\ CD,$$
$$BG,\ (AC)G,\ A(CG),\ HB,\ H^{\mathrm{T}}B,\ D^{\mathrm{T}}C,\ G^{\mathrm{T}} + F,\ G^{\mathrm{T}}D.$$

2. Let A,B be general 2×2 matrices. Write down the matrices

$$\alpha A,\ \alpha A + \beta B,\ A + B,\ AB,\ A^{\mathrm{T}},\ AA,\ AA^{\mathrm{T}},\ A + B^{\mathrm{T}}.$$

3. Let $A = [a_{ij}]^{m \times n}$. Prove that $(A^{\mathrm{T}})^{\mathrm{T}} = A$.
4. Let H be an $n \times n$ matrix. Denote HH by H^2, H^2H by H^3, and so on. For the matrix H in Exercise 1, calculate H^2, H^3, H^4, and so on.
5. Let $A = [a_{ij}]^{m \times n}$. Write down the ith row of A^{T}, and also write the jth column of A^{T}.
6. Let $A = [a_{ij}]^{m \times n}$, and let $\mathbf{r}_1, \ldots, \mathbf{r}_m$ be the rows of A, so that each $\mathbf{r}_i$ is a $1 \times n$ row vector, and

$$A = \begin{bmatrix} \mathbf{r}_1 \\ \cdot \\ \cdot \\ \cdot \\ \mathbf{r}_m \end{bmatrix}.$$

Prove that

$$A^{\mathrm{T}} = [\mathbf{r}_1^{\mathrm{T}} \ \cdots \ \mathbf{r}_m^{\mathrm{T}}].$$

7. Let $A = [a_{ij}]^{m \times n}$. Show that

$$[1 \ 0 \ 0 \ \cdots \ 0]A = [a_{11} \ \ a_{12} \ \cdots \ a_{1n}],$$
$$[0 \ 1 \ 0 \ \cdots \ 0]A = [a_{21} \ \ a_{22} \ \cdots \ a_{2n}],$$

and also show that

$$A\begin{bmatrix} 1 \\ 0 \\ \cdot \\ \cdot \\ \cdot \\ 0 \end{bmatrix} = \begin{bmatrix} a_{11} \\ a_{21} \\ \cdot \\ \cdot \\ \cdot \\ a_{m1} \end{bmatrix}, \qquad A\begin{bmatrix} 0 \\ 1 \\ 0 \\ \cdot \\ \cdot \\ \cdot \\ 0 \end{bmatrix} = \begin{bmatrix} a_{12} \\ a_{22} \\ \cdot \\ \cdot \\ \cdot \\ a_{m2} \end{bmatrix}.$$

What is the general result?

4 RULES FOR MANIPULATION

If A, B, and C are $m \times n$ matrices and α, β are scalars, then we can form various other $m \times n$ matrices, such as $A + B$, $\alpha A + \beta B$, and $(A + B) + C$. Such combinations obey rules analogous to those holding for ordinary numbers. It is easy to check that

$$A + B = B + A, \qquad (A + B) + C = A + (B + C),$$
$$\alpha A + \beta A = (\alpha + \beta)A, \qquad \alpha(-A) = -\alpha A,$$

and so on.

Multiplication is somewhat more complicated. Recall that the product XY (also written as $X \cdot Y$ for emphasis) is defined only when the number of columns in X equals the number of rows in Y. Symbolically,

$$X^{m\times n} \cdot Y^{n\times p} = (XY)^{m\times p}.$$

Whenever we write a product XY hereafter, it will always be understood that X and Y are of the right sizes so that XY is defined.

Suppose now that we are given matrices

$$A = [a_{ij}]^{m\times n}, \qquad B = [b_{ij}]^{m\times n}, \qquad C = [c_{ij}]^{n\times p}.$$

Let us prove the formula

(4.1) $$(A + B)C = AC + BC.$$

We will check that the (i,j)-entry of $(A + B)C$ equals the (i,j)-entry of $AC + BC$. Now

$$(A + B)C = \begin{bmatrix} \cdots & \cdots & & \cdots \\ a_{i1} + b_{i1} & a_{i2} + b_{i2} & \cdots & a_{in} + b_{in} \\ \cdots & \cdots & & \cdots \end{bmatrix} \begin{bmatrix} \cdots & c_{1j} & \cdots \\ \cdots & c_{2j} & \cdots \\ \cdots & \cdot & \cdots \\ \cdots & c_{nj} & \cdots \end{bmatrix},$$

where we have shown explicitly the ith row of $A + B$ and the jth column of C. Therefore the (i,j)-entry of $(A + B)C$ equals

(4.2) $$(a_{i1} + b_{i1})c_{1j} + (a_{i2} + b_{i2})c_{2j} + \cdots + (a_{in} + b_{in})c_{nj}.$$

On the other hand, the (i,j)-entry of AC is

$$a_{i1}c_{1j} + a_{i2}c_{2j} + \cdots + a_{in}c_{nj},$$

while that of BC is

$$b_{i1}c_{1j} + b_{i2}c_{2j} + \cdots + b_{in}c_{nj}.$$

The (i,j)-entry of $AC + BC$ is therefore

$$\{a_{i1}c_{1j} + a_{i2}c_{2j} + \cdots + a_{in}c_{nj}\} + \{b_{i1}c_{1j} + b_{i2}c_{2j} + \cdots + b_{in}c_{nj}\}.$$

Since this equals the expression in (4.2), for each permissible i and j, the equality (4.1) is established. [We have shown here that (4.1) is true by using the distributive law $(a + b)c = ac + bc$ for scalars.]

In the same way, we can prove that the following formulas are true, assuming always that the matrices involved are of the right sizes so that sums and products are defined:

$$(AB)C = A(BC), \qquad A \cdot \beta B = \beta A \cdot B = \beta(AB),$$
$$A(B + C) = AB + AC, \qquad A(B - C) = AB - AC.$$

Again we caution that AB and BA need not be equal, even when both products are defined. For example,

$$\begin{bmatrix} 1 & 1 \\ 0 & 1 \end{bmatrix}\begin{bmatrix} 1 & 0 \\ 1 & 1 \end{bmatrix} = \begin{bmatrix} 2 & 1 \\ 1 & 1 \end{bmatrix}, \qquad \begin{bmatrix} 1 & 0 \\ 1 & 1 \end{bmatrix}\begin{bmatrix} 1 & 1 \\ 0 & 1 \end{bmatrix} = \begin{bmatrix} 1 & 1 \\ 1 & 2 \end{bmatrix}.$$

It is also useful to have some rules for calculating with transposes of matrices. We list some of these:

$$(\alpha A)^{\mathrm{T}} = \alpha A^{\mathrm{T}}, \qquad (A + B)^{\mathrm{T}} = A^{\mathrm{T}} + B^{\mathrm{T}}, \qquad (A^{\mathrm{T}})^{\mathrm{T}} = A.$$

Somewhat less obvious is the formula,

(4.3) $$(AB)^{\mathrm{T}} = B^{\mathrm{T}}A^{\mathrm{T}}.$$

We shall prove this by showing that for each i and j, the (i,j)-entry of $(AB)^{\mathrm{T}}$ equals the (i,j)-entry of $B^{\mathrm{T}}A^{\mathrm{T}}$. Let

$$A = [a_{ij}]^{m\times n}, \qquad B = [b_{ij}]^{n\times p}.$$

The (i,j)-entry of $(AB)^{\mathrm{T}}$ equals the (j,i)-entry of AB, which is

(4.4) $$a_{j1}b_{1i} + a_{j2}b_{2i} + \cdots + a_{jn}b_{ni}.$$

The (i,j)-entry of $B^{\mathrm{T}}A^{\mathrm{T}}$ can be read off from

$$B^{\mathrm{T}}A^{\mathrm{T}} = \begin{bmatrix} \cdot & \cdot & \cdots & \cdot \\ b_{1i} & b_{2i} & \cdots & b_{ni} \\ \cdot & \cdot & \cdots & \cdot \end{bmatrix}\begin{bmatrix} \cdots & a_{j1} & \cdots \\ \cdots & a_{j2} & \cdots \\ \cdots & \cdot & \cdots \\ \cdots & a_{jn} & \cdots \end{bmatrix},$$

and is given by

$$b_{1i}a_{j1} + b_{2i}a_{j2} + \cdots + b_{ni}a_{jn}.$$

Since this expression agrees with that in (4.4), we have shown that for each i and j, the matrices $(AB)^{\mathrm{T}}$ and $B^{\mathrm{T}}A^{\mathrm{T}}$ have the same (i,j)-entry. Thus (4.3) is established.

We introduce two useful notations: a *zero matrix* is a matrix all of whose entries are zero. These zero matrices come in all sizes, and will be denoted simply by 0 if there is no danger of confusion. On the other hand, the $n \times n$ *identity matrix* I_n is given by

$$I_n = \begin{bmatrix} 1 & 0 & \cdots & 0 \\ 0 & 1 & \cdots & 0 \\ \cdot & \cdot & \cdots & \cdot \\ 0 & 0 & \cdots & 1 \end{bmatrix}^{n\times n},$$

with 1's along the main diagonal,[1] 0's elsewhere. The following identities are easily established:

$$\begin{aligned} 0 + A = A + 0 = A, &\qquad A - A = 0, \\ 0 \cdot B = 0, &\qquad B \cdot 0 = 0, \\ I_m \cdot A^{m\times n} = A^{m\times n}, &\qquad A^{m\times n} \cdot I_n = A^{m\times n}. \end{aligned}$$

Matrices of the form αI_n are called *scalar matrices*. Note that

$$\alpha I_n = \begin{bmatrix} \alpha & 0 & \cdots & 0 \\ 0 & \alpha & \cdots & 0 \\ \cdot & \cdot & \cdots & \cdot \\ 0 & 0 & \cdots & \alpha \end{bmatrix}^{n\times n}.$$

Furthermore,

$$\alpha A^{m\times n} = (\alpha I_m)A^{m\times n} = A^{m\times n}(\alpha I_n).$$

Finally, a *diagonal matrix* is a square matrix whose entries off the main diagonal are all zero. For example, a 3×3 diagonal matrix would look like

$$\begin{bmatrix} a_{11} & 0 & 0 \\ 0 & a_{22} & 0 \\ 0 & 0 & a_{33} \end{bmatrix}.$$

EXERCISES

1. Prove that $I_n^2 = I_n$.
2. If $A + B = A + C$, prove that $B = C$.
3. Let

$$A = [2 \quad -1], \qquad B = \begin{bmatrix} 1 \\ 1 \end{bmatrix}.$$

[1] For an arbitrary square matrix S, the *main diagonal* of S is the diagonal running from the upper left corner to the lower right corner of S. This main diagonal passes through the (1,1)-entry of S, the (2,2)-entry of S, and so on.

Show that $AB = I_1$. Find A^T and B^T, and check that $(AB)^T = B^T A^T$. What is $A^T B^T$?

4. Let

$$C = \begin{bmatrix} 2 & 0 \\ 1 & -1 \end{bmatrix}.$$

Calculate $3C$, C^2, $C^2 + 3C$, C^3, $C + 3I$, C^T. Find a 2×2 matrix D such that $CD = I_2$, and then calculate DC.

5. Let A and B be a pair of $n \times n$ matrices such that $AB = BA$. Prove that $AB^2 = B^2 A$, and that

$$A(B^2 + 3B + I) = (B^2 + 3B + I)A.$$

*6. For any square matrix A, and each positive integer k, show that $(A^k)^T = (A^T)^k$.

*7. If A and B are $n \times n$ matrices such that $AB = BA$, we say that A and B *commute* (with one another). Show that if A and B commute, then every power A^k commutes with every power B^l. Show also that if A commutes with B and with C, then A commutes with $B + C$.

8. Let

$$X = [x \quad y \quad z], \qquad A = \begin{bmatrix} a & b & c \\ b & d & e \\ c & e & f \end{bmatrix}.$$

Calculate XAX^T and XX^T.

9. A square matrix A is called *symmetric* if $A = A^T$. Write down a general expression for a 3×3 symmetric matrix.

10. A square matrix A is called *skew-symmetric* if $A^T = -A$. Show that the main diagonal entries of a skew-symmetric matrix are all zero. Write down general 2×2 and 3×3 skew-symmetric matrices.

11. If $B^{n \times n}$ is an arbitrary matrix, show that $B + B^T$ is symmetric, and that $B - B^T$ is skew-symmetric.

12. Prove that every square matrix B can be written as a sum of a symmetric matrix and a skew-symmetric matrix. [*Hint:* $B = \frac{1}{2}(B + B^T) + \frac{1}{2}(B - B^T)$.]

13. Let $A^{m \times n}$ be any matrix, and $\mathbf{v}$ any $n \times 1$ column vector. Prove that $(A\mathbf{v})^T = \mathbf{v}^T A^T$.

14. Let $\mathbf{v} = [c_1 \quad c_2 \quad \cdots \quad c_n]$, where each c_k is a complex number. If $c = a + bi$, a, b real, let $\bar{c} = a - bi$, and $|c|^2 = \bar{c}c = a^2 + b^2$. Show that

$$\bar{\mathbf{v}}\,\mathbf{v}^T = |c_1|^2 + \cdots + |c_n|^2,$$

where $\bar{\mathbf{v}} = [\bar{c}_1 \quad \cdots \quad \bar{c}_n]$.

5 REVIEW OF DETERMINANTS

Given a square matrix A, we associate with it a certain number called its *determinant,*[1] denoted by det A. For $A = [a]^{1\times 1}$, we set det $A = a$. For a 2×2 matrix A, we write

$$A = \begin{bmatrix} a & b \\ c & d \end{bmatrix}, \qquad \det A = \begin{vmatrix} a & b \\ c & d \end{vmatrix} = ad - bc.$$

To calculate det $A^{3\times 3}$, it is convenient to write the first two columns of A to the right of A. Let

$$A = \begin{bmatrix} a_1 & a_2 & a_3 \\ b_1 & b_2 & b_3 \\ c_1 & c_2 & c_3 \end{bmatrix}.$$

Then form the pattern

$$\begin{matrix} + & + & + & & \\ \begin{bmatrix} a_1 & a_2 & a_3 \\ b_1 & b_2 & b_3 \\ c_1 & c_2 & c_3 \end{bmatrix} & & & \begin{matrix} a_1 & a_2 \\ b_1 & b_2, \\ c_1 & c_2 \end{matrix} \\ - & - & - & & \end{matrix}$$

and write 3 products with a plus sign, as shown by arrows, and 3 products with a minus sign. This yields

$$\det A = a_1b_2c_3 + a_2b_3c_1 + a_3b_1c_2 - a_3b_2c_1 - a_1b_3c_2 - a_2b_1c_3.$$

Unfortunately, this simple procedure does not work for 4×4 or larger matrices.

We shall not give the general definition of the determinant of an $n \times n$ matrix, since that definition is seldom used in evaluating determinants. Instead, we shall give (without proof) some of the main consequences of that definition and shall show how to use them to calculate determinants. For the proofs of the theorems stated below, as well as

[1] Many authors also use the notation $|A|$ for the determinant of A.

for other omitted proofs of results stated in the later sections, we refer the reader to the references listed in the Preface.

Given a square matrix $A = [a_{ij}]^{n\times n}$, we define the *minor* (or *minor determinant*) of the element a_{ij} to be the $(n-1)\times(n-1)$ determinant obtained by crossing out the row and the column in which a_{ij} occurs. Thus if $A = [a_{ij}]^{3\times 3}$, then

$$\text{minor of } a_{11} = \begin{vmatrix} a_{22} & a_{23} \\ a_{32} & a_{33} \end{vmatrix}, \qquad \text{minor of } a_{23} = \begin{vmatrix} a_{11} & a_{12} \\ a_{31} & a_{32} \end{vmatrix},$$

and so on.

Now define the *cofactor* A_{ij} of the element a_{ij} to be either plus or minus the minor of a_{ij}, depending on whether $i+j$ is even or odd. This amounts to setting

$$A_{ij} = \text{cofactor of } a_{ij} = (-1)^{i+j}\cdot \text{minor of } a_{ij}.$$

In the example above where $A = [a_{ij}]^{3\times 3}$, we have

$$A_{11} = \begin{vmatrix} a_{22} & a_{23} \\ a_{32} & a_{33} \end{vmatrix} = a_{22}a_{33} - a_{32}a_{23},$$

$$A_{23} = -\begin{vmatrix} a_{11} & a_{12} \\ a_{31} & a_{32} \end{vmatrix} = -(a_{11}a_{32} - a_{31}a_{12}),$$

and so on.

The plus and minus signs that must be put in front of minors to give cofactors can be best remembered by the array

$$\begin{matrix} + & - & + \\ - & + & - \\ + & - & + \end{matrix}$$

(shown here for the 3×3 case). The array starts with $+$ in the upper left-hand corner, and changes sign for each sideways or downward step. Thus, for instance, the cofactor A_{22} equals the minor of a_{22}, the cofactor A_{21} is the negative of the minor of a_{21}, and so on.

When $A = [a_{ij}]^{2\times 2}$, we have

$$A = \begin{bmatrix} a_{11} & a_{12} \\ a_{21} & a_{22} \end{bmatrix},$$

$$A_{11} = a_{22}, \qquad A_{12} = -a_{21}, \qquad A_{21} = -a_{12}, \qquad A_{22} = a_{11}.$$

The following theorem is proved in the references:

Laplace Expansion Theorem

Let $A = [a_{ij}]^{n\times n}$ be a square matrix, and let A_{ij} be the cofactor of a_{ij}. Then for each fixed value of i (between 1 and n),

(5.1) $$\det A = a_{i1}A_{i1} + a_{i2}A_{i2} + \cdots + a_{in}A_{in}.$$

On the other hand, when $k \neq i$, then

(5.2) $$0 = a_{i1}A_{k1} + a_{i2}A_{k2} + \cdots + a_{in}A_{kn}.$$

We call Formula (5.1) the expansion of det A gotten by using the ith row of A. One can also form an expansion using columns:

(5.3) $$\det A = a_{1j}A_{1j} + a_{2j}A_{2j} + \cdots + a_{nj}A_{nj},$$

(5.4) $$0 = a_{1j}A_{1r} + a_{2j}A_{2r} + \cdots + a_{nj}A_{nr}, \quad \text{if } r \neq j.$$

To illustrate, let

$$A = \begin{bmatrix} a_{11} & a_{12} & a_{13} \\ a_{21} & a_{22} & a_{23} \\ a_{31} & a_{32} & a_{33} \end{bmatrix},$$

$$A_{31} = \begin{vmatrix} a_{12} & a_{13} \\ a_{22} & a_{23} \end{vmatrix}, \qquad A_{32} = -\begin{vmatrix} a_{11} & a_{13} \\ a_{21} & a_{23} \end{vmatrix}, \qquad A_{33} = \begin{vmatrix} a_{11} & a_{12} \\ a_{21} & a_{22} \end{vmatrix}.$$

Then

$$\det A = a_{31}A_{31} + a_{32}A_{32} + a_{33}A_{33} \qquad \text{(expansion along 3rd row).}$$

The Laplace Expansion Theorem enables us to reduce the calculation of an $n \times n$ determinant to the calculation of n determinants of size $(n-1) \times (n-1)$. By repeated use of this result, we can therefore evaluate determinants of any size. However, this approach is rather complicated for determinants of large size, and we shall give a simpler method below, which depends on the following results:

(5.5) Theorem[2]

Let A and B be $n \times n$ matrices, α a scalar.

(i) *If every element of a given row of A is multiplied by α, the matrix thus obtained has determinant equal to $\alpha \cdot \det A$.*
(ii) *If two rows of A are interchanged, the matrix thus obtained has determinant equal to $-\det A$.*
(iii) *If a fixed multiple*[3] *of the elements of one row of A are added to the corresponding elements of another row of A, the resulting matrix has the same determinant as A.*
(iv) *$\det(AB) = (\det A)(\det B)$.*
(v) *$\det(A^T) = \det A$.*
(vi) *The first three assertions above remain true if we use "columns" in place of "rows."*

[2] For proofs, see references in the Preface.
[3] The multiplier may be negative, of course.

Let us illustrate parts (i)–(iii) for 2×2 matrices. By (i) we obtain

$$\det \begin{bmatrix} a_1 & a_2 \\ \alpha b_1 & \alpha b_2 \end{bmatrix} = \alpha \cdot \det \begin{bmatrix} a_1 & a_2 \\ b_1 & b_2 \end{bmatrix}.$$

From (ii) we get

$$\det \begin{bmatrix} a_1 & a_2 \\ b_1 & b_2 \end{bmatrix} = -\det \begin{bmatrix} b_1 & b_2 \\ a_1 & a_2 \end{bmatrix},$$

while (iii) gives

$$\det \begin{bmatrix} a_1 - tb_1 & a_2 - tb_2 \\ b_1 & b_2 \end{bmatrix} = \det \begin{bmatrix} a_1 & a_2 \\ b_1 & b_2 \end{bmatrix}.$$

(In this last equation, we subtracted t times row 2 from row 1.)

We now give a simple method for calculating $\det A$. To begin, subtract multiples of one row of A from all the other rows of A, in such a manner as to get a column having only one nonzero term. Then use the Laplace expansion along that column.

ILLUSTRATION

$$A = \begin{bmatrix} 2 & 1 & 5 & 0 \\ 1 & 3 & 1 & 1 \\ 4 & -2 & 2 & -1 \\ 3 & 1 & -5 & 1 \end{bmatrix} \xrightarrow[r_1 - 2r_2]{\text{step 1}} \begin{bmatrix} 0 & -5 & 3 & -2 \\ 1 & 3 & 1 & 1 \\ 4 & -2 & 2 & -1 \\ 3 & 1 & -5 & 1 \end{bmatrix}$$

$$\xrightarrow[r_3 - 4r_2]{\text{step 2}} \begin{bmatrix} 0 & -5 & 3 & -2 \\ 1 & 3 & 1 & 1 \\ 0 & -14 & -2 & -5 \\ 3 & 1 & -5 & 1 \end{bmatrix} \xrightarrow[r_4 - 3r_2]{\text{step 3}} \begin{bmatrix} 0 & -5 & 3 & -2 \\ 1 & 3 & 1 & 1 \\ 0 & -14 & -2 & -5 \\ 0 & -8 & -8 & -2 \end{bmatrix}.$$

In step 1, we subtracted $2 \cdot$ row 2 from row 1; in step 2, we subtracted $4 \cdot$ row 2 from row 3; in step 3, we subtracted $3 \cdot$ row 2 from row 4. Obviously, we could have done all these steps at once and just written down the final matrix.

Now use the Laplace expansion along the first column of the final matrix, and remember that the cofactor of the element in the (2,1)-position is the negative of the minor of that element. We then get

$$\det A = -(1) \cdot \begin{vmatrix} -5 & 3 & -2 \\ -14 & -2 & -5 \\ -8 & -8 & -2 \end{vmatrix}.$$

Now by part (i) of Theorem 5.5, changing the sign of each element in the second row of the above 3×3 determinant has the effect of changing the sign of the determinant. Another sign change arises when we change the sign of each element in the third row. Therefore

$$\det A = -\begin{vmatrix} -5 & 3 & -2 \\ 14 & 2 & 5 \\ 8 & 8 & 2 \end{vmatrix}.$$

But now we can repeat the procedure to obtain

$$\begin{bmatrix} -5 & 3 & -2 \\ 14 & 2 & 5 \\ 8 & 8 & 2 \end{bmatrix} \xrightarrow[r_1 + r_3,\ r_2 - \frac{5}{2} r_3]{\text{step 4}} \begin{bmatrix} 3 & 11 & 0 \\ -6 & -18 & 0 \\ 8 & 8 & 2 \end{bmatrix},$$

where we added row 3 to row 1, and subtracted $\frac{5}{2}$ · row 3 from row 2. Using the Laplace expansion along the third column, we get

$$\begin{vmatrix} 3 & 11 & 0 \\ -6 & -18 & 0 \\ 8 & 8 & 2 \end{vmatrix} = 2\begin{vmatrix} 3 & 11 \\ -6 & -18 \end{vmatrix} = 2(-54 + 66) = 24,$$

and so $\det A = -24$.

This procedure is in practice a very efficient method for calculating large size determinants.

EXERCISES

1. Let

$$A = \begin{bmatrix} 1 & 2 & 3 \\ 4 & 5 & -1 \\ 0 & 2 & 3 \end{bmatrix}.$$

Find the cofactors of the elements of A. Then use formula (5.1) with $i = 1$, and again with $i = 3$, to calculate $\det A$. Also evaluate $\det A$ by the method described at the end of this section.

2. Evaluate each of the determinants

$$\begin{vmatrix} 1 & 3 & 1 & -1 \\ 0 & 2 & 4 & 1 \\ -1 & 1 & 2 & 0 \\ 0 & 3 & 1 & 3 \end{vmatrix}, \quad \begin{vmatrix} -1 & 1 & 0 & 3 \\ 3 & 0 & 1 & 1 \\ 2 & -1 & 2 & 2 \\ 2 & 3 & 0 & 1 \end{vmatrix}, \quad \begin{vmatrix} 1 & -1 & 1 \\ -1 & 1 & 1 \\ 1 & 1 & 1 \end{vmatrix}.$$

3. Prove that

$$\begin{vmatrix} a & b & c \\ a' & b' & c' \\ a'' & b'' & c'' \end{vmatrix} = 0$$

if there exist scalars α, β such that

$$a = \alpha a' + \beta a'', \qquad b = \alpha b' + \beta b'', \qquad c = \alpha c' + \beta c''.$$

4. What is $\det(\alpha I_n)$, where α is a scalar?
5. Let $A = [a_{ij}]^{2\times 2}$, $D = [A_{ij}]^{2\times 2}$, where A_{ij} is the *cofactor* of a_{ij}.

Prove that

$$AD^{\mathrm{T}} = D^{\mathrm{T}}A = \alpha I, \quad \text{where } \alpha = \det A.$$

6. Let $A = [a_{ij}]^{4\times 4}$ be an *upper triangular matrix,* that is, all of its entries below the main diagonal are zero. Prove that

$$\det A = a_{11}a_{22}a_{33}a_{44}.$$

7. Let

$$A = \begin{bmatrix} B^{2\times 2} & 0^{2\times 3} \\ 0^{3\times 2} & C^{3\times 3} \end{bmatrix} = \begin{bmatrix} b_{11} & b_{12} & 0 & 0 & 0 \\ b_{21} & b_{22} & 0 & 0 & 0 \\ 0 & 0 & c_{11} & c_{12} & c_{13} \\ 0 & 0 & c_{21} & c_{22} & c_{23} \\ 0 & 0 & c_{31} & c_{32} & c_{33} \end{bmatrix}.$$

Show that $\det A = (\det B)(\det C)$.

8. Let

$$A = \begin{bmatrix} B^{2\times 2} & D^{2\times 3} \\ 0^{3\times 2} & C^{3\times 3} \end{bmatrix}.$$

Show that $\det A = (\det B)(\det C)$.

9. A square matrix A is called *skew-symmetric* if $A^{\mathrm{T}} = -A$. Show that the main diagonal entries of such a matrix are all zero. Calculate the determinant of a general 2×2 and 3×3 skew-symmetric matrix.

*10. Let $P_1(x_1,y_1)$ and $P_2(x_2,y_2)$ be distinct points in the XY-plane. Show that the equation of the line through P_1 and P_2 is given by

$$\begin{vmatrix} x & y & 1 \\ x_1 & y_1 & 1 \\ x_2 & y_2 & 1 \end{vmatrix} = 0.$$

[*Hint:* Prove that this is a linear equation, not identically zero, satisfied by (x_1,y_1) and (x_2,y_2). Also, see Remark on page 22.]

*11. Let P_1, P_2, P_3 be three noncollinear points in XYZ-space, and let P_i have coordinates (x_i,y_i,z_i), $i = 1,2,3$. Show that the equation of the plane through P_1, P_2, and P_3 is

$$\begin{vmatrix} x & y & z & 1 \\ x_1 & y_1 & z_1 & 1 \\ x_2 & y_2 & z_2 & 1 \\ x_3 & y_3 & z_3 & 1 \end{vmatrix} = 0.$$

(*Caution:* the difficulty lies in showing that in the expansion of the above determinant, the coefficients of x, y, and z are not all zero. Also, see Remark on page 22.)

12. For any square matrix A and any positive integer k, prove that

$$\det(A^k) = (\det A)^k.$$

13. If A and B are square matrices such that $AB = I$, prove that $\det A \neq 0$.

Remark: In connection with Exercises 10 and 11, we state without proof the following results (i) and (ii) from analytic geometry:

(i) Let P_1, P_2, P_3 be points in the XY-plane, and let (x_i,y_i) denote the coordinates of P_i, $i = 1,2,3$. Then

$$\begin{vmatrix} x_1 & y_1 & 1 \\ x_2 & y_2 & 1 \\ x_3 & y_3 & 1 \end{vmatrix} = \pm 2 \cdot \text{Area of triangle } P_1P_2P_3,$$

with the $+$ occurring if P_1, P_2, P_3 occur in counterclockwise order, and the $-$ otherwise.

(ii) Let P_1, P_2, P_3, P_4 be points in XYZ-space, and let P_i have coordinates (x_i,y_i,z_i), $i = 1,2,3,4$. Then

$$\begin{vmatrix} x_1 & y_1 & z_1 & 1 \\ x_2 & y_2 & z_2 & 1 \\ x_3 & y_3 & z_3 & 1 \\ x_4 & y_4 & z_4 & 1 \end{vmatrix} = \pm 6 \cdot \text{Volume of tetrahedron } P_1P_2P_3P_4.$$

The $+$ or $-$ depends on the geometric configuration of the vertices (see Figure 5.1).

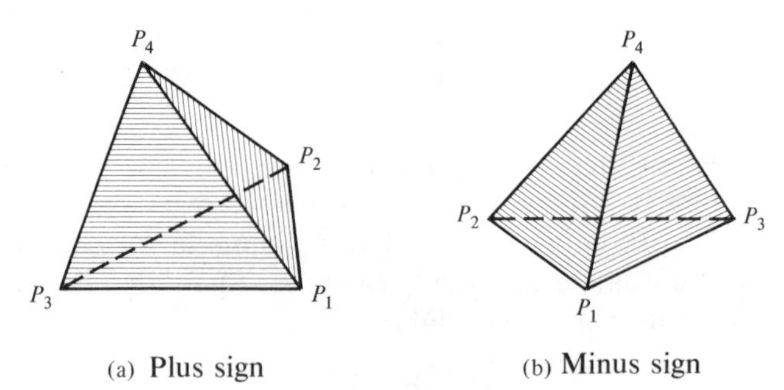

(a) Plus sign (b) Minus sign

Figure 5.1

(iii) Let us show how the first remark can be used to solve Exercise 10. Suppose the points P_1, P_2 are given as in that exercise, and let $P(x,y)$ denote an arbitrary point in the XY-plane. According to the remark, the expression

(5.6)

$$\begin{vmatrix} x & y & 1 \\ x_1 & y_1 & 1 \\ x_2 & y_2 & 1 \end{vmatrix}$$

equals $\pm 2 \cdot$ Area of triangle PP_1P_2. Hence this expression equals zero if and only if the triangle PP_1P_2 is degenerate and lies along a straight line. But this means that the expression (5.6) equals zero if and only if the point P lies on the line through P_1 and P_2. Therefore, the equation of line P_1P_2 is obtained by setting the expression (5.6) equal to zero. Note that this equation is in fact a linear equation in the variables x, y.

A corresponding discussion can be given to solve Exercise 11.

6 INVERSES OF MATRICES

Let A and B be $n \times n$ matrices such that

$$AB = BA = I_n.$$

We call A and B *inverses* of one another, and write $B = A^{-1}$, $A = B^{-1}$. (Read A^{-1} as "*A inverse.*") Note that

$$A = B^{-1} = (A^{-1})^{-1}.$$

We shall say that the matrix A is *nonsingular* (or *invertible*) if A has an inverse.

Only a square matrix can have an inverse. Some square matrices do not have inverses, however, and are called *singular* matrices. For instance, the 3×3 zero matrix has no inverse, and is singular.

We remark that if A has an inverse, it has only one inverse. For suppose that both B and C are inverses of A. Then B and C must be equal, since

$$B = BI = B(AC) = (BA)C = IC = C.$$

If A has an inverse A^{-1}, then $AA^{-1} = I$, and so

$$(\det A)(\det A^{-1}) = \det(AA^{-1}) = \det I = 1.$$

Thus if A is nonsingular, then $\det A \neq 0$. We shall see that the converse is also true:

If $\det A \neq 0$, then A has an inverse.

In order to prove this important fact, we introduce the concept of the *adjoint* of a square matrix A, denoted by adj A. This is defined as follows: replace each entry of A by its cofactor, and then take the transpose of this matrix of cofactors; this transpose matrix is adj A.

EXAMPLES

1. Let $A = \begin{bmatrix} a & b \\ c & d \end{bmatrix}$. The cofactor of a is d, the cofactor of b is $-c$, the

cofactor of c is $-b$, and the cofactor of d is a. The matrix of cofactors is therefore $\begin{bmatrix} d & -c \\ -b & a \end{bmatrix}$, and so adj $A = \begin{bmatrix} d & -b \\ -c & a \end{bmatrix}$.

2. Let $A = [a_{ij}]^{3\times 3}$, $A_{ij} =$ cofactor of a_{ij}. Then

$$\operatorname{adj} A = \begin{bmatrix} A_{11} & A_{21} & A_{31} \\ A_{12} & A_{22} & A_{32} \\ A_{13} & A_{23} & A_{33} \end{bmatrix} = \text{transpose of the matrix of cofactors.}$$

(6.1) Theorem

Let A be an $n \times n$ matrix for which $\alpha = \det A \neq 0$. Then

$$A^{-1} = \frac{1}{\alpha} \cdot \operatorname{adj} A,$$

where the right-hand expression is the scalar $1/\alpha$ times the matrix adj A.

Before proving the theorem, let us give two illustrations.

(i) If $ad - bc \neq 0$, then

$$\begin{bmatrix} a & b \\ c & d \end{bmatrix}^{-1} = \frac{1}{ad - bc} \begin{bmatrix} d & -b \\ -c & a \end{bmatrix}.$$

(ii) Let

$$A = \begin{bmatrix} 0 & 2 & 2 \\ -1 & 3 & 2 \\ 1 & 0 & 5 \end{bmatrix}.$$

The cofactors of the entries of A are given by

$$A_{11} = \begin{vmatrix} 3 & 2 \\ 0 & 5 \end{vmatrix} = 15, \quad A_{12} = -\begin{vmatrix} -1 & 2 \\ 1 & 5 \end{vmatrix} = 7, \quad A_{13} = \begin{vmatrix} -1 & 3 \\ 1 & 0 \end{vmatrix} = -3,$$

$$A_{21} = -\begin{vmatrix} 2 & 2 \\ 0 & 5 \end{vmatrix} = -10, \quad A_{22} = \begin{vmatrix} 0 & 2 \\ 1 & 5 \end{vmatrix} = -2, \quad A_{23} = -\begin{vmatrix} 0 & 2 \\ 1 & 0 \end{vmatrix} = 2,$$

$$A_{31} = \begin{vmatrix} 2 & 2 \\ 3 & 2 \end{vmatrix} = -2, \quad A_{32} = -\begin{vmatrix} 0 & 2 \\ -1 & 2 \end{vmatrix} = -2, \quad A_{33} = \begin{vmatrix} 0 & 2 \\ -1 & 3 \end{vmatrix} = 2.$$

Thus

$$\operatorname{adj} A = \begin{bmatrix} 15 & -10 & -2 \\ 7 & -2 & -2 \\ -3 & 2 & 2 \end{bmatrix}, \qquad \begin{aligned} \det A &= a_{11}A_{11} + a_{12}A_{12} + a_{13}A_{13} \\ &= 2 \cdot 7 + 2(-3) = 8. \end{aligned}$$

Therefore

$$A^{-1} = \tfrac{1}{8} \begin{bmatrix} 15 & -10 & -2 \\ 7 & -2 & -2 \\ -3 & 2 & 2 \end{bmatrix} = \begin{bmatrix} \frac{15}{8} & \frac{-10}{8} & \frac{-2}{8} \\ \frac{7}{8} & \frac{-2}{8} & \frac{-2}{8} \\ \frac{-3}{8} & \frac{2}{8} & \frac{2}{8} \end{bmatrix}.$$

(The reader should verify that the above 3×3 matrix is indeed the inverse of A by checking that the equalities $A \cdot A^{-1} = I$ and $A^{-1} \cdot A = I$ hold true.)

Now let us prove Theorem 6.1. We must show that

$$A \cdot \alpha^{-1}(\text{adj } A) = (\alpha^{-1}(\text{adj } A)) \cdot A = I.$$

We have

$$A \cdot \alpha^{-1}(\text{adj } A) = \begin{bmatrix} \cdot & \cdot & \cdots & \cdot \\ a_{i1} & a_{i2} & \cdots & a_{in} \\ \cdot & \cdot & \cdots & \cdot \end{bmatrix} \begin{bmatrix} \cdot & \alpha^{-1}A_{j1} & \cdot \\ \cdot & \alpha^{-1}A_{j2} & \cdot \\ \cdot & \cdot & \cdot \\ \cdot & \cdot & \cdot \\ \cdot & \cdot & \cdot \\ \cdot & \alpha^{-1}A_{jn} & \cdot \end{bmatrix},$$

where we have indicated the ith row of the first factor, and the jth column of the second factor. Therefore, the (i,j)-entry of $A \cdot \alpha^{-1}(\text{adj } A)$ is

$$\alpha^{-1}(a_{i1}A_{j1} + a_{i2}A_{j2} + \cdots + a_{in}A_{jn}) = \begin{cases} 1, & \text{if } j = i \\ 0, & \text{if } j \neq i, \end{cases}$$

by using identities (5.1) and (5.2). But the (i,j)-entry of I_n is 1 if $j = i$, and 0 if $j \neq i$. Therefore $A \cdot \alpha^{-1} \text{ adj } A = I$. The formula $(\alpha^{-1} \text{ adj } A)A = I$ is established in a similar manner, and so (6.1) is proved.

As a consequence of Theorem 6.1, we can now assert:

(6.2) *Let A be a square matrix. Then A has an inverse if and only if $\det A \neq 0$. In other words, A is nonsingular if and only if $\det A \neq 0$.*

(6.3) *Remark:* Let A and D be square matrices such that $AD = I$. Then we claim that A must have an inverse A^{-1}, and that $D = A^{-1}$. Indeed, if $AD = I$ then $(\det A)(\det D) = \det I = 1$, so $\det A \neq 0$. By (6.1) it follows that A^{-1} exists. But then

$$D = ID = (A^{-1}A)D = A^{-1}(AD) = A^{-1}I = A^{-1},$$

as claimed.

If A is nonsingular, we may deduce from the equation $AX = C$ that $X = A^{-1}C$; in fact,

$$X = (A^{-1}A)X = A^{-1}(AX) = A^{-1}C.$$

Likewise, if $YA = F$, then $Y = FA^{-1}$. The following rules are useful, and we leave their proofs to the exercises:

$$\textbf{(6.4)}\qquad \begin{aligned}(A^{-1})^{-1} &= A \\ (AB)^{-1} &= B^{-1}A^{-1}, \\ (A^{\mathrm{T}})^{-1} &= (A^{-1})^{\mathrm{T}}, \\ (A^k)^{-1} &= (A^{-1})^k.\end{aligned} \qquad \begin{aligned} I^{-1} &= I, \\ (ABC)^{-1} &= C^{-1}B^{-1}A^{-1}, \\ (\lambda A)^{-1} &= \lambda^{-1}A^{-1},\end{aligned}$$

In these formulas, A, B, and C are arbitrary nonsingular matrices, λ is a nonzero scalar, and k is a nonzero integer.

To conclude this section, we shall describe another way of calculating inverses of nonsingular matrices, which does not involve using adjoints, and which is more efficient for large size matrices. Given a nonsingular square matrix $A^{n\times n}$, proceed as follows: pick any row of A with a nonzero first entry and subtract suitable multiples of that row from the other rows of A, so as to make all other entries in the first column equal to 0. Then choose *another* row of this new matrix with a nonzero second entry and kill off the other second column entries by row subtractions, and so on. After this process has been carried out n times, we then permute rows of the resulting matrix to get a diagonal matrix, say with main diagonal entries $x_1, \ldots, x_n$. Now, $\det A = \pm x_1 \cdots x_n$, so since $\det A \neq 0$, then also each $x_i \neq 0$. Finally, multiply the first row by x_1^{-1}, the second row by x_2^{-1}, and so on. *If the same set of operations is performed on the rows of the identity matrix I_n, the final matrix obtained will be precisely A^{-1}.*

Before attempting to justify the method, let us illustrate it by applying it to the 3×3 matrix A considered in (ii) following (6.1). We have

$$A = \begin{bmatrix} 0 & 2 & 2 \\ -1 & 3 & 2 \\ 1 & 0 & 5 \end{bmatrix} \xrightarrow[r_2 + r_3]{①} \begin{bmatrix} 0 & 2 & 2 \\ 0 & 3 & 7 \\ 1 & 0 & 5 \end{bmatrix} \xrightarrow[r_2 - \frac{3}{2}r_1]{②} \begin{bmatrix} 0 & 2 & 2 \\ 0 & 0 & 4 \\ 1 & 0 & 5 \end{bmatrix}$$

$$\xrightarrow[r_1 - \frac{1}{2}r_2,\ r_3 - \frac{5}{4}r_2]{③} \begin{bmatrix} 0 & 2 & 0 \\ 0 & 0 & 4 \\ 1 & 0 & 0 \end{bmatrix} \xrightarrow[r_3 \to r_1]{④} \begin{bmatrix} 1 & 0 & 0 \\ 0 & 2 & 0 \\ 0 & 0 & 4 \end{bmatrix} \xrightarrow{⑤} \begin{bmatrix} 1 & 0 & 0 \\ 0 & 1 & 0 \\ 0 & 0 & 1 \end{bmatrix}.$$

In step 1, we add row 3 to row 2; in step 2, subtract $\frac{3}{2}$ · row 1 from row 2; in step 3, subtract $\frac{1}{2}$ · row 2 from row 1 and $\frac{5}{4}$ · row 2 from row 3; in step 4, move row 3 up to the first position; in step 5, multiply row 2 by $\frac{1}{2}$, row 3 by $\frac{1}{4}$. Now apply these operations to the rows of I_3, getting

$$I = \begin{bmatrix} 1 & 0 & 0 \\ 0 & 1 & 0 \\ 0 & 0 & 1 \end{bmatrix} \xrightarrow{①} \begin{bmatrix} 1 & 0 & 0 \\ 0 & 1 & 1 \\ 0 & 0 & 1 \end{bmatrix} \xrightarrow{②} \begin{bmatrix} 1 & 0 & 0 \\ -\frac{3}{2} & 1 & 1 \\ 0 & 0 & 1 \end{bmatrix} \xrightarrow{③} \begin{bmatrix} \frac{7}{4} & -\frac{1}{2} & -\frac{1}{2} \\ -\frac{3}{2} & 1 & 1 \\ \frac{15}{8} & -\frac{5}{4} & -\frac{1}{4} \end{bmatrix}$$

$$\xrightarrow{④} \begin{bmatrix} \frac{15}{8} & -\frac{5}{4} & -\frac{1}{4} \\ \frac{7}{4} & -\frac{1}{2} & -\frac{1}{2} \\ -\frac{3}{2} & 1 & 1 \end{bmatrix} \xrightarrow{⑤} \begin{bmatrix} \frac{15}{8} & -\frac{5}{4} & -\frac{1}{4} \\ \frac{7}{8} & -\frac{1}{4} & -\frac{1}{4} \\ -\frac{3}{8} & \frac{1}{4} & \frac{1}{4} \end{bmatrix} = A^{-1}.$$

*[Why does this work? We give a condensed sketch of the justification of this procedure: note first that

$$\begin{bmatrix} 1 & t & 0 \\ 0 & 1 & 0 \\ 0 & 0 & 1 \end{bmatrix} \begin{bmatrix} a & a' & a'' \\ b & b' & b'' \\ c & c' & c'' \end{bmatrix} = \begin{bmatrix} a+tb & a'+tb' & a''+tb'' \\ b & b' & b'' \\ c & c' & c'' \end{bmatrix}.$$

This shows that multiplication by

$$M = \begin{bmatrix} 1 & t & 0 \\ 0 & 1 & 0 \\ 0 & 0 & 1 \end{bmatrix}$$

on the left has the effect of adding to the first row of

$$\begin{bmatrix} a & a' & a'' \\ b & b' & b'' \\ c & c' & c'' \end{bmatrix}$$

t times its second row. It is not hard to show that each of the operations described above corresponds in a similar fashion to left multiplication by some suitable matrix. Assuming this fact, the proof of which we leave as an exercise for the reader, it follows that our procedure amounts to determining a set of matrices $M_1, M_2, \ldots, M_k$ such that

$$M_k \cdots M_2 M_1 A = I.$$

(Here, M_1 corresponds to the operation performed in the first step, M_2 to the second step, and so on.) By (6.3) we may then conclude that

$$A^{-1} = M_k \cdots M_2 M_1.$$

But on the other hand, the result of successively applying these same operations to the rows of I is precisely the matrix

$$M_k \cdots M_2 M_1 \cdot I,$$

which is the same as $M_k \cdots M_2 M_1$, that is, A^{-1}. This brief sketch may help you understand why the procedure described above is valid.]*

EXERCISES

1. Find inverses of each of the following matrices, if such inverses exist:

$$[5], \begin{bmatrix} 2 & -1 \\ 3 & 2 \end{bmatrix}, \begin{bmatrix} 1 & x \\ 0 & 1 \end{bmatrix}, \begin{bmatrix} 1 & 0 \\ y & 1 \end{bmatrix}, \begin{bmatrix} a & 0 \\ 0 & b \end{bmatrix}, \begin{bmatrix} a & -b \\ b & a \end{bmatrix},$$

$$\begin{bmatrix} 2 & 1 & 3 \\ -1 & 0 & 4 \end{bmatrix}, \begin{bmatrix} 2 & 1 & 3 \\ -1 & 0 & 4 \\ 1 & 3 & 0 \end{bmatrix}, \begin{bmatrix} 2 & 3 & 0 & 0 \\ 1 & 1 & 0 & 0 \\ 0 & 0 & 3 & 1 \\ 0 & 0 & 2 & 1 \end{bmatrix}.$$

2. Find the adjoints of

$$\begin{bmatrix} 2 & 3 \\ -1 & 4 \end{bmatrix}, \begin{bmatrix} 6 & 1 & 0 \\ 2 & -1 & 1 \\ 0 & 2 & 3 \end{bmatrix}.$$

In Exercises 3–9, let A and B be nonsingular $n \times n$ matrices.

3. Take transposes in the equations $AA^{-1} = A^{-1}A = I$ (using 4.3) to get $(A^{-1})^{\mathrm{T}}A^{\mathrm{T}} = A^{\mathrm{T}}(A^{-1})^{\mathrm{T}} = I$. Deduce from this that $(A^{\mathrm{T}})^{-1} = (A^{-1})^{\mathrm{T}}$.
4. If $AXB = C$, show that $X = A^{-1}CB^{-1}$.
5. Prove that $(AB)^{-1} = B^{-1}A^{-1}$, by checking that

$$(AB)(B^{-1}A^{-1}) = I,$$

and then using (6.3).
6. Prove that $(A^{-1}B)^{-1} = B^{-1}A$.
7. Prove that $(A^2)^{-1} = (A^{-1})^2$. [*Hint:* Take $B = A$ in Exercise 5.]
*8. Prove that for any positive integer k,

$$(AXA^{-1})(AYA^{-1}) = A \cdot XY \cdot A^{-1}, \qquad (AXA^{-1})^k = AX^kA^{-1},$$

and

$$(A^k)^{-1} = (A^{-1})^k.$$

9. Prove that $\{(AB)^{-1}\}^{\mathrm{T}} = (A^{-1})^{\mathrm{T}} \cdot (B^{-1})^{\mathrm{T}}$.
10. For what values of x is the matrix $\begin{bmatrix} x-3 & 4 \\ 2 & x-1 \end{bmatrix}$ singular?
*11. Let A be any $n \times n$ matrix, possibly singular, and let $\alpha = \det A$. Using the method of proof of Theorem 6.1, show that

$$A \cdot \operatorname{adj} A = \alpha I_n.$$

By taking determinants of both sides, prove that

$$(\det A)(\det(\operatorname{adj} A)) = \alpha^n.$$

If $\alpha \neq 0$, deduce from this that

$$\det(\operatorname{adj} A) = \alpha^{n-1}.$$

As a matter of fact, this last result holds true even when $\alpha = 0$; verify it directly for the case when $n = 2$.
12. If A and B are nonsingular $n \times n$ matrices, is it necessarily true that AB and $A + B$ are also nonsingular?
13. If $A^{n \times n}$ is singular, and $X^{n \times n}$ is arbitrary, show that the product AX is also singular.

7 LINEAR EQUATIONS, CRAMER'S RULE

Consider a system of n simultaneous linear equations in the n unknowns $x_1, \ldots, x_n$:

$$\text{(7.1)} \qquad \begin{cases} a_{11}x_1 + a_{12}x_2 + \cdots + a_{1n}x_n = b_1 \\ a_{21}x_1 + a_{22}x_2 + \cdots + a_{2n}x_n = b_2 \\ \quad \cdots \\ a_{n1}x_1 + a_{n2}x_2 + \cdots + a_{nn}x_n = b_n. \end{cases}$$

Set $A = [a_{ij}]^{n \times n} =$ *matrix of coefficients* of the unknowns. Let us arrange the unknowns to form a column vector. From the definition of matrix multiplication, we have

$$A \cdot \begin{bmatrix} x_1 \\ x_2 \\ \cdot \\ \cdot \\ \cdot \\ x_n \end{bmatrix} = \begin{bmatrix} a_{11} & \cdots & a_{1n} \\ a_{21} & \cdots & a_{2n} \\ & \cdots & \\ a_{n1} & \cdots & a_{nn} \end{bmatrix} \begin{bmatrix} x_1 \\ x_2 \\ \cdot \\ \cdot \\ \cdot \\ x_n \end{bmatrix} = \begin{bmatrix} a_{11}x_1 + a_{12}x_2 + \cdots + a_{1n}x_n \\ a_{21}x_1 + a_{22}x_2 + \cdots + a_{2n}x_n \\ \cdots \\ a_{n1}x_1 + a_{n2}x_2 + \cdots + a_{nn}x_n \end{bmatrix}.$$

Therefore equations (7.1) may be rewritten more compactly as

$$\text{(7.2)} \qquad A \cdot \begin{bmatrix} x_1 \\ x_2 \\ \vdots \\ x_n \end{bmatrix} = \begin{bmatrix} b_1 \\ b_2 \\ \vdots \\ b_n \end{bmatrix}.$$

Suppose now that $\det A \neq 0$, so that the matrix A is nonsingular, and therefore there exists a unique matrix A^{-1} such that $AA^{-1} = A^{-1}A = I$. If $x_1, \ldots, x_n$ satisfy equations (7.2) [and thus also (7.1)], then

$$\text{(7.3)} \qquad \begin{bmatrix} x_1 \\ \cdot \\ \cdot \\ \cdot \\ x_n \end{bmatrix} = A^{-1}A \begin{bmatrix} x_1 \\ \cdot \\ \cdot \\ \cdot \\ x_n \end{bmatrix} = A^{-1} \begin{bmatrix} b_1 \\ \cdot \\ \cdot \\ \cdot \\ b_n \end{bmatrix}.$$

Thus, if there is a solution to (7.2), it must be unique. Conversely, if $x_1, \ldots, x_n$ are given in terms of $b_1, \ldots, b_n$ by (7.3), then these x's are indeed a solution of equations (7.2), since

$$A\begin{bmatrix} x_1 \\ \cdot \\ \cdot \\ \cdot \\ x_n \end{bmatrix} = A \cdot A^{-1}\begin{bmatrix} b_1 \\ \cdot \\ \cdot \\ \cdot \\ b_n \end{bmatrix} = \begin{bmatrix} b_1 \\ \cdot \\ \cdot \\ \cdot \\ b_n \end{bmatrix}.$$

We have therefore shown

(7.4) Theorem

If $\det A \neq 0$, *then the system of simultaneous linear equations* (7.1) *has a unique solution* $x_1, \ldots, x_n$, *given by*

$$\begin{bmatrix} x_1 \\ \cdot \\ \cdot \\ \cdot \\ x_n \end{bmatrix} = A^{-1}\begin{bmatrix} b_1 \\ \cdot \\ \cdot \\ \cdot \\ b_n \end{bmatrix}.$$

An immediate consequence of this result is the following useful fact:

(7.5) Corollary

Let $A = [a_{ij}]^{n \times n}$, *and suppose that* $\det A \neq 0$. *Then the only possible solution of the system of homogeneous linear equations*

$$\text{(7.6)} \qquad \begin{cases} a_{11}x_1 + \cdots + a_{1n}x_n = 0 \\ \qquad \cdots \\ a_{n1}x_1 + \cdots + a_{nn}x_n = 0 \end{cases}$$

is the "trivial solution" $x_1 = 0, \ldots, x_n = 0$.

PROOF: We apply Theorem 7.4 with $b_1 = 0, \ldots, b_n = 0$. Then

$$\begin{bmatrix} x_1 \\ \cdot \\ \cdot \\ \cdot \\ x_n \end{bmatrix} = A^{-1}\begin{bmatrix} b_1 \\ \cdot \\ \cdot \\ \cdot \\ b_n \end{bmatrix} = \begin{bmatrix} 0 \\ \cdot \\ \cdot \\ \cdot \\ 0 \end{bmatrix},$$

so each x_i must be zero.

In practice, we often use the "contrapositive" of (7.5), namely

(7.7) Theorem

If equations (7.6) have a nontrivial solution (that is, a solution in which at least one of the unknowns $x_1, \ldots, x_n$ is different from zero), then necessarily $\det A = 0$.

PROOF: If $\det A \neq 0$, then by (7.5) the equations (7.6) have only the trivial solution. Thus if there is a nontrivial solution, we must have $\det A = 0$.

We may also derive from Theorem 7.4 the following basic result:

(7.8) Cramer's Rule

Let A be the matrix of coefficients in the system of linear equations (7.1), and suppose that $\det A \neq 0$. For each i (from 1 to n), let M_i be the matrix obtained from A by replacing the ith column of A by the column of constant terms $b_1, \ldots, b_n$, leaving the other columns of A unchanged. Then the system (7.1) has the unique solution

$$x_1 = \frac{\det M_1}{\det A}, \qquad x_2 = \frac{\det M_2}{\det A}, \quad \ldots, \qquad x_n = \frac{\det M_n}{\det A}.$$

Before proving Cramer's Rule, let us illustrate it. Consider the simultaneous equations

$$\begin{cases} 2x + 3y - 4z = 7 \\ x + y + 3z = -1, \\ 4x \qquad - z = 6 \end{cases} \text{and set } A = \begin{bmatrix} 2 & 3 & -4 \\ 1 & 1 & 3 \\ 4 & 0 & -1 \end{bmatrix}.$$

Then $\det A \neq 0$, and the unique solution of the system of equations is

$$x = \frac{\begin{vmatrix} 7 & 3 & 4 \\ -1 & 1 & 3 \\ 6 & 0 & -1 \end{vmatrix}}{\det A}, \qquad y = \frac{\begin{vmatrix} 2 & 7 & -4 \\ 1 & -1 & 3 \\ 4 & 6 & -1 \end{vmatrix}}{\det A}, \qquad z = \frac{\begin{vmatrix} 2 & 3 & 7 \\ 1 & 1 & -1 \\ 4 & 0 & 6 \end{vmatrix}}{\det A}.$$

*[We shall now prove Cramer's Rule. Since we are assuming that $\det A \neq 0$, we may apply Theorem 7.4 to obtain the unique solution of equations (7.1); this solution is given by

$$\begin{bmatrix} x_1 \\ \cdot \\ \cdot \\ \cdot \\ x_n \end{bmatrix} = A^{-1} \begin{bmatrix} b_1 \\ \cdot \\ \cdot \\ \cdot \\ b_n \end{bmatrix}.$$

Set $\alpha = \det A$. According to Theorem 6.1, we may express A^{-1} in terms of α^{-1} and the adjoint of A, as follows:

$$A^{-1} = \alpha^{-1} \cdot \text{adj } A = \alpha^{-1} \begin{bmatrix} A_{11} & \cdots & A_{n1} \\ & \cdots & \\ A_{1n} & \cdots & A_{nn} \end{bmatrix},$$

where A_{ij} denotes the cofactor of the entry a_{ij} of the matrix A. Therefore

$$\begin{bmatrix} x_1 \\ \cdot \\ \cdot \\ \cdot \\ x_n \end{bmatrix} = \alpha^{-1} \begin{bmatrix} A_{11} & \cdots & A_{n1} \\ & \cdots & \\ A_{1n} & \cdots & A_{nn} \end{bmatrix} \begin{bmatrix} b_1 \\ \cdot \\ \cdot \\ \cdot \\ b_n \end{bmatrix}$$

$$= \alpha^{-1} \begin{bmatrix} A_{11}b_1 + A_{21}b_2 + \cdots + A_{n1}b_n \\ \cdots \\ A_{1n}b_1 + A_{2n}b_2 + \cdots + A_{nn}b_n \end{bmatrix}.$$

We may then conclude that for each i,

(7.9) $$x_i = \alpha^{-1}(A_{1i}b_1 + A_{2i}b_2 + \cdots + A_{ni}b_n).$$

Once we prove that

(7.10) $$\det M_i = A_{1i}b_1 + A_{2i}b_2 + \cdots + A_{ni}b_n,$$

we will be able to rewrite (7.9) as

$$x_i = \alpha^{-1} \det M_i = \frac{\det M_i}{\det A},$$

and the proof will be completed.

We are thus left with the task of proving that (7.10) is true. By definition, the matrix M_i is given by

$$M_i = \begin{bmatrix} a_{11} & \cdots & a_{1,i-1} & b_1 & a_{1,i+1} & \cdots & a_{1n} \\ a_{21} & \cdots & a_{2,i-1} & b_2 & a_{2,i+1} & \cdots & a_{2n} \\ & \cdots & & & & \cdots & \\ a_{n1} & \cdots & a_{n,i-1} & b_n & a_{n,i+1} & \cdots & a_{nn} \end{bmatrix},$$

since M_i is obtained from A by replacing its ith column by the column of constant terms $b_1, b_2, \ldots, b_n$. The cofactor of the entry b_1 of M_1 is $\pm$ minor of b_1, and this minor is the $(n-1) \times (n-1)$ determinant gotten by crossing out the elements in the first row of M_i and the ith column of M_i. Therefore

$$\text{cofactor of } b_1 = \pm \begin{vmatrix} a_{21} & \cdots & a_{2,i-1} & a_{2,i+1} & \cdots & a_{2n} \\ & \cdots & & & \cdots & \\ a_{n1} & \cdots & a_{n,i-1} & a_{n,i+1} & \cdots & a_{nn} \end{vmatrix}.$$

But the right-hand expression above is precisely the same as the cofactor A_{1i} of the element a_{1i} of A. Likewise, the cofactor of b_2 in M_i equals A_{2i}, and so on. Now use the Laplace expansion of det M_i along the ith column of M_i, obtaining

$$\det M_i = (\text{cofactor of } b_1) \cdot b_1 + (\text{cofactor of } b_2) \cdot b_2 + \cdots + (\text{cofactor of } b_n) \cdot b_n$$
$$= A_{1i}b_1 + A_{2i}b_2 + \cdots + A_{ni}b_n.$$

This establishes (7.10), and completes the proof of Cramer's Rule.]*

It is not always convenient to use Cramer's Rule for solving simultaneous linear equations, especially those with many variables, since the computation of the determinants involved can be tedious. We therefore give another method, which can in fact be applied to any number of equations in any number of unknowns. Suppose we want to solve the system of m simultaneous linear equations in n unknowns $x_1, \ldots, x_n$:

(7.11)
$$\begin{cases} a_{11}x_1 + a_{12}x_2 + \cdots + a_{1n}x_n = b_1 \\ a_{21}x_2 + a_{22}x_2 + \cdots + a_{2n}x_n = b_2 \\ \cdots \\ a_{m1}x_1 + a_{m2}x_2 + \cdots + a_{mn}x_n = b_m. \end{cases}$$

We proceed by successive "elimination of variables," as follows: pick any equation in which x_1 occurs, say the first equation for convenience, and subtract suitable multiples of the first equation from each of the others, so as to eliminate the x_1 term in each of the other equations. We then have a new set of equations with precisely the same solutions as the original set, but where x_1 now occurs only in the first equation. Solve that equation for x_1 in terms of $x_2, \ldots, x_n$, and then go on to the remaining $m - 1$ equations in the remaining $n - 1$ unknowns $x_2, \ldots, x_n$. For this set of $m - 1$ equations we repeat the procedure: pick an equation in which, let us assume, x_2 occurs, and eliminate x_2 from all of the other equations, and so on.

In this elimination procedure, it may happen that some of the new equations which arise will have only zero on the left-hand side; if the corresponding constant on the right-hand side is also zero, we can disregard the equation, since it is just the identity $0 = 0$. On the other hand, if the corresponding constant term is *not* zero, then the original system is *inconsistent,* and *has no solution.*

Some examples may help clarify the method. Note that after the elimination procedure is completed, we solve the resulting system of equations "from the bottom up." This means that we use the last equation to solve for one of the variables in terms of the remaining variables; then substitute for that variable in the next to the last equation, and solve that equation for another variable, and so on. (See Example 4 following, for instance.)

EXAMPLES

1. $\begin{cases} x - y = 5 \\ 2x + y = 4 \end{cases} \longrightarrow \begin{cases} x - y = 5 \\ 3y = -6 \end{cases} \longrightarrow \begin{cases} x = 5 + y \\ y = -2 \end{cases} \longrightarrow \begin{cases} x = 3 \\ y = -2. \end{cases}$

2. $\begin{cases} x - y = 5 \\ 2x - 2y = 4 \end{cases} \longrightarrow \begin{cases} x - y = 5 \\ 0 = -6 \end{cases} \longrightarrow$ no solution.

3. $\begin{cases} x - y = 5 \\ 2x - 2y = 10 \end{cases} \longrightarrow \begin{cases} x - y = 5 \\ 0 = 0 \end{cases} \longrightarrow \begin{cases} x = 5 + y \\ y \text{ arbitrary.} \end{cases}$

4. $\begin{cases} x + y + z = 3 \\ 2x + z = 4 \\ 2y + z = 2 \end{cases} \longrightarrow \begin{cases} x + y + z = 3 \\ -2y - z = -2 \\ 2y + z = 2 \end{cases} \longrightarrow \begin{cases} x + y + z = 3 \\ -2y - z = -2 \\ 0 = 0 \end{cases}$

$$\longrightarrow \begin{cases} x = 3 - y - z = 3 - y - (2 - 2y) \\ z = 2 - 2y \\ 0 = 0 \end{cases} \longrightarrow \begin{cases} x = 1 + y \\ z = 2 - 2y \\ y \text{ arbitrary.} \end{cases}$$

5. $\begin{cases} x + 2y - z - 3w + u = 4 \\ 2x + 3y + z + 2u = 10 \\ y + 2w + u = 5 \end{cases} \longrightarrow \begin{cases} x + 2y - z - 3w + u = 4 \\ -y + 3z + 6w = 2 \\ y + 2w + u = 5 \end{cases}$

$$\longrightarrow \begin{cases} x + 2y - z - 3w + u = 4 \\ -y + 3z + 6w = 2 \\ 3z + 8w + u = 7 \end{cases}$$

$$\longrightarrow \begin{cases} x = 4 - 2y + z + 3w - u = \frac{1}{3}(-11 + 2u + 13w) \\ y = 3z + 6w - 2 = 5 - u - 2w \\ z = \frac{1}{3}(7 - u - 8w) \\ u, w \text{ arbitrary.} \end{cases}$$

In Example 5, we have solved for x, y, z in terms of u, w; the variables u and w may have any values, and once these are chosen, x, y, z are determined.

Consider again the system (7.11) of m equations in the n unknowns $x_1, \ldots, x_n$. If the system is consistent, the method of "eliminating variables" just described will lead us to a new set of m equations. In this new set, suppose that $m - r$ of the equations are the identity $0 = 0$. Each of the other r equations can be used to solve for one of the variables x_k, for example, in terms of the variables $x_{k+1}, \ldots, x_n$. Therefore in the final solution, exactly r of the variables $x_1, \ldots, x_n$ will be expressed in terms of the remaining $n - r$ variables, which may take arbitrary values. It is clear that $r \leqslant m$ and $r \leqslant n$.

We shall call r the *rank* of the matrix of coefficients $[a_{ij}]^{m \times n}$. This rank depends only upon the matrix $[a_{ij}]^{m \times n}$, and not upon the manner in which we successively eliminate variables.[1] (Intuitively speaking, in

[1] For proof, see references in the Preface.

a consistent system of equations in n variables, the rank r is the number of "independent" constraints placed on these variables. Thus $n - r$ of the variables are still free to vary, and the remaining r variables are uniquely determined once these $n - r$ variables are fixed.)

Of particular importance is the case of n homogeneous equations in n unknowns:

(7.12)
$$\begin{cases} a_{11}x_1 + \cdots + a_{1n}x_n = 0 \\ \cdots \\ a_{n1}x_1 + \cdots + a_{nn}x_n = 0. \end{cases}$$

This system is surely consistent, since there is always the "trivial" solution $x_1 = 0, \ldots, x_n = 0$.

(7.13) Theorem

Let $A = [a_{ij}]^{n \times n}$ be the matrix of coefficients of a system of n homogeneous linear equations in n unknowns. Then there exists a nontrivial solution if and only if $\det A = 0$.

*[*PROOF:* In the procedure for eliminating variables, each step has the effect of subtracting from one row of the matrix of coefficients some multiple of another row. By Theorem 5.5, this does not change the determinant of the matrix of coefficients. The final system of equations will look like this:

(7.14)
$$\begin{cases} a'_{11}x_1 + a'_{12}x_2 + a'_{13}x_3 + \cdots + a'_{1n}x_n = 0 \\ \qquad a'_{22}x_2 + a'_{23}x_3 + \cdots + a'_{2n}x_n = 0 \\ \qquad\qquad a'_{33}x_3 + \cdots + a'_{3n}x_n = 0 \\ \qquad\qquad\qquad \cdots \\ \qquad\qquad\qquad\qquad a'_{nn}x_n = 0. \end{cases}$$

Then by Exercise 5.6,

$$\det A = \det \begin{bmatrix} a'_{11} & a'_{12} & \cdots & a'_{1n} \\ 0 & a'_{22} & \cdots & a'_{2n} \\ & & \cdots & \\ 0 & 0 & & a'_{nn} \end{bmatrix} = a'_{11}a'_{22} \cdots a'_{nn}.$$

If $\det A \neq 0$, then each $a'_{ii} \neq 0$, so if we solve equations (7.14) from the bottom up, we get only the trivial solution $x_n = 0, x_{n-1} = 0, \ldots, x_1 = 0$. On the other hand, if $\det A = 0$, then some $a'_{ii} = 0$. Suppose for example that $a'_{11} \neq 0$, $a'_{22} \neq 0$, but $a'_{33} = 0$. Then we can obtain a nontrivial solution by choosing $x_3 = 1, x_4 = 0, \ldots, x_n = 0$, and solving the first two equations in (7.14) for x_2 and x_1. This completes the proof.]*

(7.15) Corollary

Let $A = [a_{ij}]^{n \times n}$. Then A has rank n if and only if $\det A \neq 0$.

*[*PROOF:* Let r be the rank of A; we have already seen that $r \leqslant n$. According to the discussion preceding (7.13), in the system of simul-

taneous equations (7.12), we may solve for r of the variables in terms of the remaining $n-r$ variables, which may be arbitrary. If $r < n$, this shows that (7.12) has nontrivial solutions, so $\det A = 0$ by (7.13). On the other hand, if $r = n$ then (7.12) has a unique solution (the trivial one), so by (7.13) we get $\det A \neq 0$. Together these show that $r = n$ if and only if $\det A \neq 0$, and the result is proved.]*

EXERCISES

1. Solve by Cramer's Rule:

$$\begin{cases} 2x + 3y = 7 \\ x + 2y = 3, \end{cases} \qquad \begin{cases} x + y - z = 2 \\ 2x + z = 3 \\ x - y + 4z = -1 \end{cases}$$

2. Solve each of the following systems of simultaneous equations by successive elimination of variables:

$$\begin{cases} 3x + y + z = 1 \\ 6x - y = 7 \\ 6y + z = 5, \end{cases} \qquad \begin{cases} x + y + z = 1 \\ 3x + 3y + z = 5, \end{cases} \qquad \begin{cases} x + y + z = 1 \\ y + z = -1, \end{cases}$$

$$\begin{cases} x + y - z = 2 \\ -x + y + z = 4 \\ x - y + z = 6, \end{cases} \qquad \begin{cases} x + y = 9 \\ 2x + y = 3 \\ x - y = 4, \end{cases} \qquad \begin{cases} x + y + z = -4 \\ 2x + 2y + 2z = -8. \end{cases}$$

3. Solve

$$\begin{cases} x + y + 3z - w = 2 \\ y + w = 5, \end{cases} \qquad \begin{cases} x - y + w + z = 10 \\ y - z = 4 \\ x + w = 14. \end{cases}$$

4. Let u and v be given by

$$u = ax + by, \qquad v = cx + dy,$$

where a, b, c, and d are constants such that $ad - bc \neq 0$. Prove that

$$x = \frac{1}{ad - bc}(du - bv), \qquad y = \frac{1}{ad - bc}(-cu + av).$$

5. Let $A = [a_{ij}]^{3 \times 3}$ be a nonsingular matrix, and let

$$\begin{aligned} a_{11}x + a_{12}y + a_{13}z &= u \\ a_{21}x + a_{22}y + a_{23}z &= v \\ a_{31}x + a_{32}y + a_{33}z &= w. \end{aligned}$$

Prove that

$$\begin{bmatrix} x \\ y \\ z \end{bmatrix} = A^{-1} \begin{bmatrix} u \\ v \\ w \end{bmatrix}.$$

6. Let $A = [a_{ij}]^{2\times 2}$ be nonsingular, and suppose that

$$A \cdot \begin{bmatrix} x \\ y \end{bmatrix} = A \cdot \begin{bmatrix} x' \\ y' \end{bmatrix}.$$

Prove that $x = x'$ and $y = y'$.

7. Let $A^{3\times 3}$ be nonsingular, and suppose that at least one of x, y, z is nonzero. Let

$$[x' \quad y' \quad z'] = [x \quad y \quad z] \cdot A.$$

Show that at least one of x', y', z' is not zero.

8. Let $A^{n\times n}$ be a square matrix such that $A\mathbf{v} = \mathbf{0}$ for some nonzero $n \times 1$ vector $\mathbf{v}$. Show that $\det A = 0$. Prove conversely that if $\det A = 0$, then there exists a nonzero vector $\mathbf{v}$ for which $A\mathbf{v} = \mathbf{0}$.

*9. Consider the system of 3 homogeneous equations in 4 unknowns:

$$\begin{cases} a_1x_1 + a_2x_2 + a_3x_3 + a_4x_4 = 0 \\ b_1x_1 + b_2x_2 + b_3x_3 + b_4x_4 = 0 \\ c_1x_1 + c_2x_2 + c_3x_3 + c_4x_4 = 0, \end{cases}$$

and suppose that the matrix

$$A = \begin{bmatrix} a_1 & a_2 & a_3 \\ b_1 & b_2 & b_3 \\ c_1 & c_2 & c_3 \end{bmatrix}$$

is nonsingular. Use Cramer's Rule to prove that

$$x_1 = \frac{-\begin{vmatrix} a_4 & a_2 & a_3 \\ b_4 & b_2 & b_3 \\ c_4 & c_2 & c_3 \end{vmatrix}}{\det A} x_4, \qquad x_2 = \frac{-\begin{vmatrix} a_1 & a_4 & a_3 \\ b_1 & b_4 & b_3 \\ c_1 & c_4 & c_3 \end{vmatrix}}{\det A} x_4, \qquad x_3 = \frac{-\begin{vmatrix} a_1 & a_2 & a_4 \\ b_1 & b_2 & b_4 \\ c_1 & c_2 & c_4 \end{vmatrix}}{\det A} x_4.$$

Show that this may be written as

$$\frac{x_1}{\begin{vmatrix} a_2 & a_3 & a_4 \\ b_2 & b_3 & b_4 \\ c_2 & c_3 & c_4 \end{vmatrix}} = \frac{-x_2}{\begin{vmatrix} a_1 & a_3 & a_4 \\ b_1 & b_3 & b_4 \\ c_1 & c_3 & c_4 \end{vmatrix}} = \frac{x_3}{\begin{vmatrix} a_1 & a_2 & a_4 \\ b_1 & b_2 & b_4 \\ c_1 & c_2 & c_4 \end{vmatrix}} = \frac{-x_4}{\begin{vmatrix} a_1 & a_2 & a_3 \\ b_1 & b_2 & b_3 \\ c_1 & c_2 & c_3 \end{vmatrix}}.$$

(If one of the denominators is zero, the corresponding x is also zero.) This formula gives the ratios $x_1 : -x_2 : x_3 : -x_4$ provided that there is at least one nonzero 3×3 minor from the 3×4 matrix of coefficients.

8 LINEAR TRANSFORMATIONS

Let P denote a point with coordinates (x,y) in the XY-plane. We introduce two new variables u, v by setting

(8.1) $$u = ax + by, \qquad v = cx + dy, \quad a,b,c,d \text{ constants.}$$

Let Q be the point with coordinates (u,v) in the UV-plane. By means of equations (8.1), each point P in the XY-plane determines a point Q in the UV-plane. For example, if $P = (0,0)$, then $Q = (0,0)$, while if $P = (1,0)$, then $Q = (a,c)$.

We shall say that equations (8.1) define a *linear transformation* of the XY-plane into the UV-plane, and that this transformation *maps* the point P onto the point Q.

Generally speaking, as P ranges over all points on some curve C in the XY-plane, the corresponding point Q traces out some curve C' in the UV-plane. In particular, suppose that C is a straight line through (0,0) in the XY-plane. Let us show that (apart from exceptional cases) the corresponding C' is a straight line through (0,0) in the UV-plane. For this purpose it is most convenient to describe C by parametric equations, as follows. Let (α,β) denote some fixed point on C other than (0,0). Then the parametric equations of C can be written as

$$x = \alpha t, \qquad y = \beta t, \qquad t = \text{parameter.}$$

As t ranges over all real numbers, the point (x,y) determined by these parametric equations ranges over all points of the line C. Substituting for x and y in equations (8.1), we see that the points on C' have coordinates (u,v), where

$$\begin{cases} u = a \cdot \alpha t + b \cdot \beta t = (a\alpha + b\beta)t, \\ v = c \cdot \alpha t + d \cdot \beta t = (c\alpha + d\beta)t, \end{cases} \qquad t = \text{parameter.}$$

Thus C' is a straight line through (0,0) and $(a\alpha + b\beta,\ c\alpha + d\beta)$ in the UV-plane, apart from the exceptional case in which both $a\alpha + b\beta = 0$ and $c\alpha + d\beta = 0$; in this exceptional case, C' consists of the single point (0,0). We may remark that since α and β are not both zero, it follows from (7.5) that the exceptional case does not arise when $ad - bc \neq 0$. Thus if we assume that $ad - bc \neq 0$, then the transformation (8.1) maps each straight line through the origin in the XY-plane onto a straight line through the origin in the UV-plane. For this reason we call it a *linear* transformation.

It is convenient to interpret the linear transformation in another way, as follows: it maps the vector $\overrightarrow{OP}$ in the XY-plane onto the vector $\overrightarrow{OQ}$ in the UV-plane. (See Figure 8.1.)

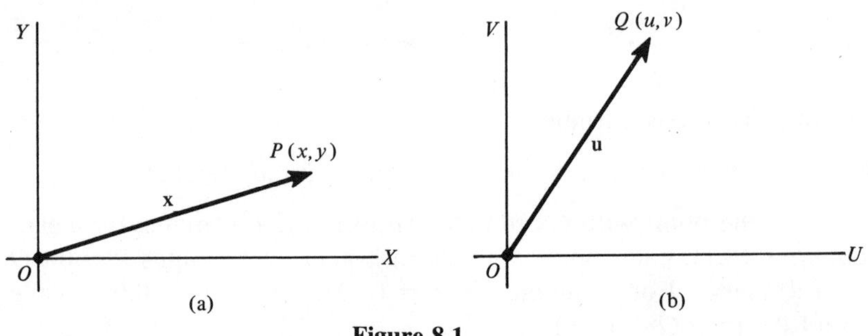

Figure 8.1

Let us represent $\overrightarrow{OP}$ and $\overrightarrow{OQ}$ as column vectors, and put

$$\mathbf{x} = \overrightarrow{OP} = \begin{bmatrix} x \\ y \end{bmatrix}, \qquad \mathbf{u} = \overrightarrow{OQ} = \begin{bmatrix} u \\ v \end{bmatrix}, \qquad A = \begin{bmatrix} a & b \\ c & d \end{bmatrix}.$$

We may then rewrite (8.1) as

$$\begin{bmatrix} a & b \\ c & d \end{bmatrix} \begin{bmatrix} x \\ y \end{bmatrix} = \begin{bmatrix} u \\ v \end{bmatrix}, \qquad \text{that is, } A\mathbf{x} = \mathbf{u}.$$

Thus the linear transformation defined by the matrix A maps the vector $\mathbf{x}$ in the XY-plane onto the vector $A\mathbf{x}$ in the UV-plane.

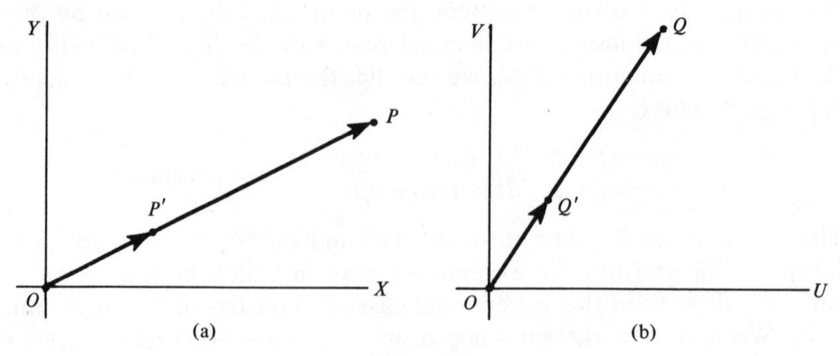

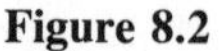

Figure 8.2

Let us show next that each point P' on the line segment OP is mapped onto a point Q' on the line segment OQ, in such a manner that if $OP' = \alpha \cdot OP$, then $OQ' = \alpha \cdot OQ$. (See Figure 8.2.) Clearly $\vec{OP'} = \alpha \cdot \vec{OP}$, and so

$$\vec{OQ'} = A \cdot \vec{OP'} = A(\alpha\mathbf{x}) = \alpha(A\mathbf{x}) = \alpha \cdot \vec{OQ}.$$

Thus Q' is on the line segment OQ, and $OQ' = \alpha \cdot OQ$, as claimed.

Occasionally it is desirable to give the following geometrical interpretation of the linear transformation (8.1). We superimpose the UV-plane on the XY-plane, so that the U-axis coincides with the X-axis, and the V-axis with the Y-axis. Then we can think of the transformation (8.1) as moving the point P with coordinates (x,y) in the XY-plane to the new position Q whose X- and Y-coordinates are $(ax + by, cx + dy)$. To emphasize the fact that the transformation maps the point (x,y) onto the point $(ax + by, cx + dy)$, we sometimes write it as

$$(x,y) \longrightarrow (ax + by, cx + dy).$$

For example, the transformation

$$(x,y) \longrightarrow (2x,y)$$

moves each point horizontally so as to double its displacement from the Y-axis. (See Figure 8.3.)

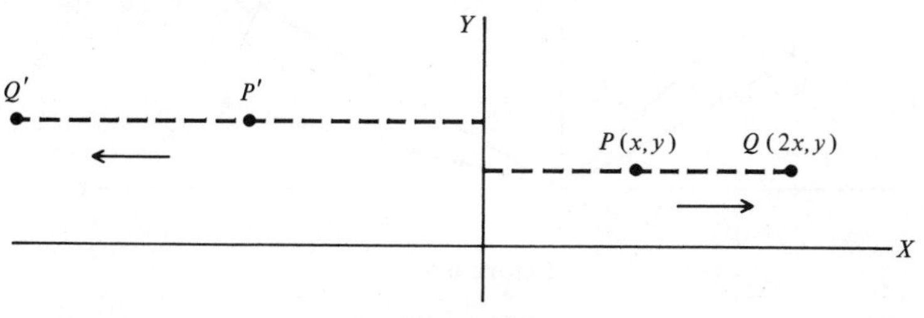

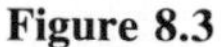

Figure 8.3

Likewise, the linear transformation

$$(x,y) \longrightarrow (y,x)$$

maps each point P onto its reflection Q across the line $y = x$. (See Figure 8.4.)

Finally, consider the linear transformation

$$P = (x,y) \longrightarrow Q = (x \cos \alpha - y \sin \alpha,\ x \sin \alpha + y \cos \alpha).$$

Let r, θ be polar coordinates of the point P, so that

$$x = r \cos \theta, \qquad y = r \sin \theta.$$

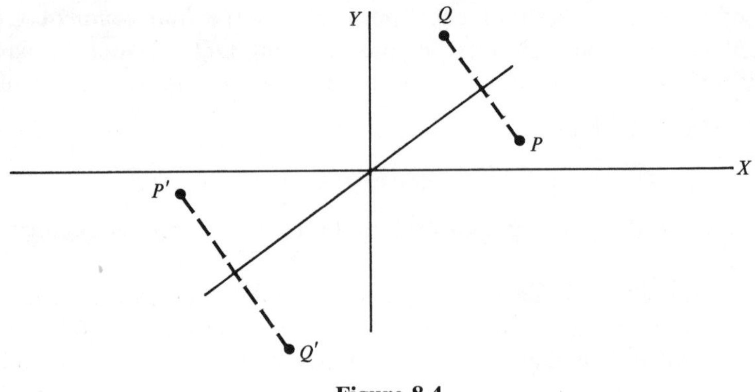

Figure 8.4

Then the rectangular coordinates of Q are

$$(r \cos \theta \cos \alpha - r \sin \theta \sin \alpha,\ r \cos \theta \sin \alpha + r \sin \theta \cos \alpha),$$

that is, $(r \cos(\theta + \alpha),\ r \sin(\theta + \alpha))$. Thus Q has polar coordinates $r, \theta + \alpha$. This means that the transformation is a rotation about the origin through an angle α; this rotation carries the point P onto the point Q. (See Figure 8.5.)

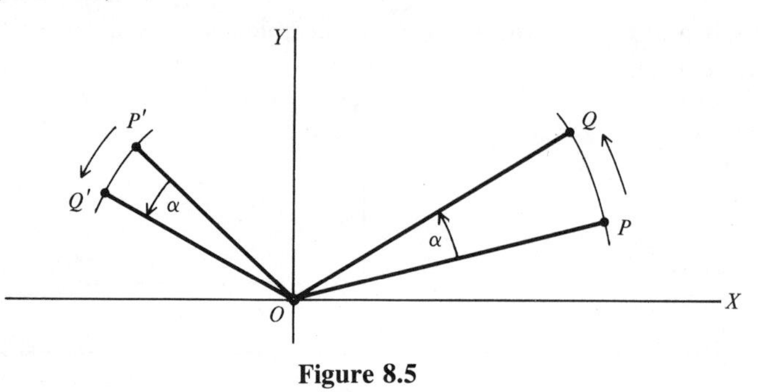

Figure 8.5

Let us now consider the especially simple linear transformation

$$u = ax, \qquad v = by, \qquad a > 0, \qquad b > 0.$$

This transformation stretches all horizontal distances by a factor of a, and all vertical distances by a factor of b. Hence any square in the XY-plane with sides parallel to the axes has its area magnified by a factor of ab in passing over to the UV-plane. (See Figure 8.6.) Suppose next that this transformation maps the region inside the curve C in the XY-plane onto the region inside the corresponding curve C' in the UV-plane. We may define the area inside C by counting small squares and using limits, as follows: for each positive integer n, draw a grid-

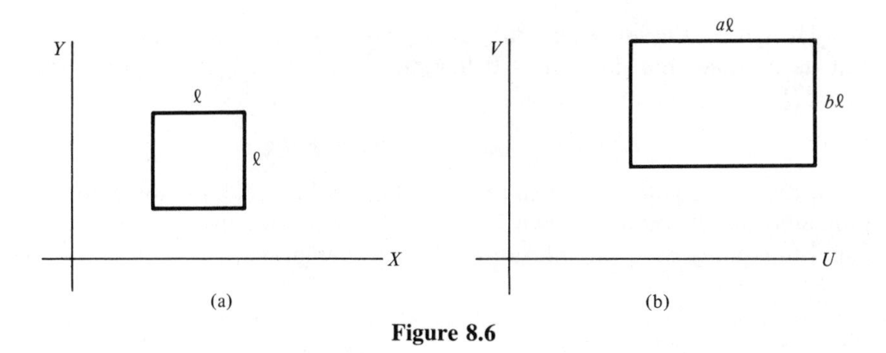

Figure 8.6

work of lines across C parallel to the X- and Y-axes, with separation $1/n$. (See Figure 8.7.) If there are $f(n)$ small squares inside C, each with area $1/n^2$, define the area inside C to be $\lim_{n\to\infty} f(n)\cdot\frac{1}{n^2}$.

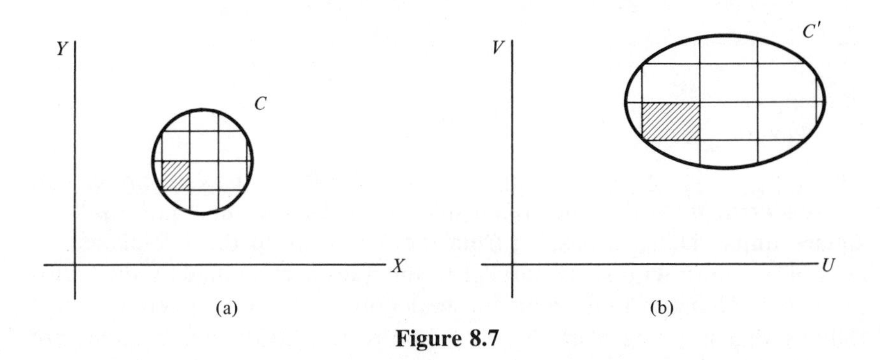

Figure 8.7

Each square in the XY-plane maps onto a rectangle in the UV-plane, so inside C' there are $f(n)$ rectangles, each of area ab/n^2. It is then intuitively clear that the area inside C' is $\lim_{n\to\infty} f(n)\cdot\frac{ab}{n^2}$, which is the same as $ab\cdot\lim_{n\to\infty} f(n)\cdot\frac{1}{n^2}$. Thus, in passing from the XY-plane to the UV-plane, *all* areas are magnified by the factor ab.

We may give a beautiful application of the above discussion. The area of the circular region $x^2+y^2\leqslant 1$ is π square units. Now $x=u/a$, $y=v/b$, so the circular region in the XY-plane is mapped onto the elliptical region in the UV-plane defined by

$$\frac{u^2}{a^2}+\frac{v^2}{b^2}\leqslant 1.$$

Therefore the area of this elliptical region is πab square units!

How does the more general linear transformation (8.1) affect areas? Let us assume that $ad - bc \neq 0$ hereafter. Let $P_0(x_0,y_0)$ map onto the point $Q_0(u_0,v_0)$, so that

$$u_0 = ax_0 + by_0, \qquad v_0 = cx_0 + dy_0.$$

Then the $l \times l$ square shown in the XY-plane in Figure 8.8 maps onto the indicated parallelogram shown in the UV-plane, with the interior of the square mapping onto the interior of the parallelogram.

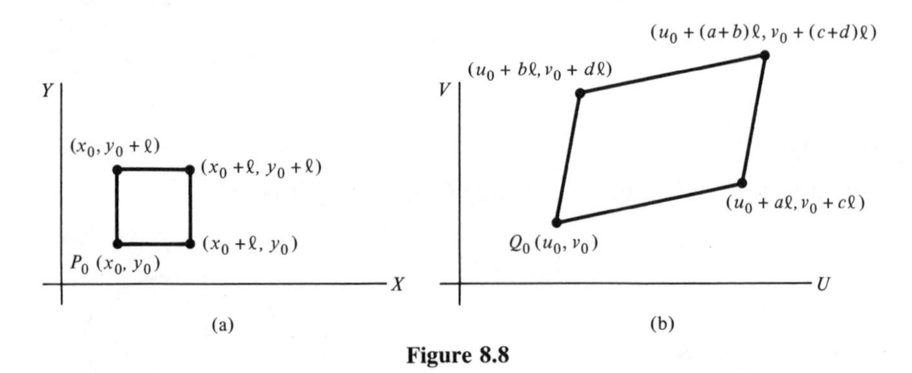

Figure 8.8

The area of the square is l^2 square units. We leave it as an exercise to the reader to show that the area inside the parallelogram is $|ad - bc| \cdot l^2$ square units. Thus, in passing from the XY-plane to the UV-plane, the area of a square with sides parallel to the axes is magnified by the factor $|ad - bc|$. Using a limit argument analogous to the one given above, it follows that in going from the XY-plane to the UV-plane, *all* areas are magnified by the factor $|ad - bc|$. Of course, we cannot help noticing that

$$ad - bc = \det \begin{bmatrix} a & b \\ c & d \end{bmatrix}.$$

This can hardly be a coincidence, and is in fact a special case of a more general result to be discussed in more detail in Section 15, when we return to the question of how linear (and nonlinear) transformations affect areas and volumes.

EXERCISES

*1. Consider the linear transformation (8.1), where $ad - bc \neq 0$. Show that every line C' through the origin in the UV-plane corresponds to some line C through the origin in the XY-plane.

2. Given the linear transformation

$$u = 2x + 2y, \qquad v = x + y,$$

which lines through the origin in the XY-plane are mapped onto a single point in the UV-plane?

*3. Given the linear transformation (8.1) in which $ad - bc \neq 0$, find out what line in the UV-plane corresponds to the line $Ax + By + C = 0$ in the XY-plane.

4. Describe geometrically the effect of the transformation

$$u = 5 + 4x, \qquad v = 1 + 2y.$$

*5. Describe geometrically the effect of the transformation

$$u = u_0 + ax + by, \qquad v = v_0 + cx + dy, \qquad ad - bc \neq 0.$$

(This is called an *affine* transformation; it is *not* a linear transformation unless both u_0 and v_0 are zero.)

6. What is the volume of the region inside the ellipsoid

$$\frac{u^2}{a^2} + \frac{v^2}{b^2} + \frac{w^2}{c^2} = 1?$$

7. Find a linear transformation of the XY-plane into itself which carries each point P onto its reflection across the X-axis.

8. Describe geometrically the linear transformation

$$(x,y) \longrightarrow (x + 3y,\ y)$$

of the XY-plane into itself.

*9. Let L be the line through the origin of the XY-plane with angle of inclination α. Show that the linear transformation of the XY-plane into itself given by

$$(x,y) \longrightarrow (x \cos 2\alpha + y \sin 2\alpha,\ x \sin 2\alpha - y \cos 2\alpha)$$

maps each point P onto its reflection across the line L.

9 VECTOR SPACES

We begin with a number of basic definitions. Let $\mathbf{v}_1, \ldots, \mathbf{v}_k$ be vectors in some n-dimensional space (for example, each $\mathbf{v}_i$ might be a $1 \times n$ row vector), and let $\alpha_1, \ldots, \alpha_k$ be scalars. The vector $\alpha_1\mathbf{v}_1 + \cdots + \alpha_k\mathbf{v}_k$ is called a *linear combination* of $\mathbf{v}_1, \ldots, \mathbf{v}_k$. The collection of *all* such linear combinations of $\mathbf{v}_1, \ldots, \mathbf{v}_k$ is called the *vector space spanned by the set of vectors* $\{\mathbf{v}_1, \ldots, \mathbf{v}_k\}$.

For example, let $\mathbf{v}_1$ be a unit vector along the X-axis, $\mathbf{v}_2$ along the Y-axis, in the 3-dimensional XYZ-space. We claim that the vector space spanned by $\{\mathbf{v}_1, \mathbf{v}_2\}$ consists of all vectors $\overrightarrow{OP}$, where P is an arbitrary point of the XY-plane. To see this, note that $\alpha\mathbf{v}_1 + \beta\mathbf{v}_2 = \overrightarrow{OP}$, where P has coordinates $(\alpha,\beta,0)$; conversely, if P is a point in the XY-plane with coordinates $(a,b,0)$, then $\overrightarrow{OP} = a\mathbf{v}_1 + b\mathbf{v}_2$.

We have thus seen, in the special case above, that the vector space spanned by $\mathbf{v}_1$ and $\mathbf{v}_2$ consists of all vectors lying in the plane of $\mathbf{v}_1$ and $\mathbf{v}_2$. A corresponding geometrical interpretation holds true for any vector space.

EXAMPLES

1. The vector space spanned by a single vector $\mathbf{v}_1$ consists of all scalar multiples $\alpha\mathbf{v}_1$ of the vector $\mathbf{v}_1$.

2. Let

$$\mathbf{v}_1 = (1 \quad 0 \quad 0 \quad 0), \qquad \mathbf{v}_2 = (0 \quad 1 \quad 0 \quad 0),$$
$$\mathbf{v}_3 = (0 \quad 0 \quad 1 \quad 0), \qquad \mathbf{v}_4 = (0 \quad 0 \quad 0 \quad 1).$$

The vector space V spanned by the set $\{\mathbf{v}_1, \mathbf{v}_2, \mathbf{v}_3, \mathbf{v}_4\}$ consists of all linear combinations $\alpha_1\mathbf{v}_1 + \alpha_2\mathbf{v}_2 + \alpha_3\mathbf{v}_3 + \alpha_4\mathbf{v}_4$, where $\alpha_1, \alpha_2, \alpha_3, \alpha_4$ are arbitrary scalars. Since

$$\alpha_1\mathbf{v}_1 + \alpha_2\mathbf{v}_2 + \alpha_3\mathbf{v}_3 + \alpha_4\mathbf{v}_4 = (\alpha_1 \quad \alpha_2 \quad \alpha_3 \quad \alpha_4),$$

it is clear that V consists of *all* 1×4 row vectors.

3. Let $A = \begin{bmatrix} 3 & 1 & 1 \\ 0 & 1 & 1 \end{bmatrix}$. The row vectors of A are

$$\mathbf{r}_1 = [3 \quad 1 \quad 1], \qquad \mathbf{r}_2 = [0 \quad 1 \quad 1].$$

The vector space spanned by $\mathbf{r}_1$ and $\mathbf{r}_2$ consists of all linear combinations $\alpha_1\mathbf{r}_1 + \alpha_2\mathbf{r}_2$; this vector space is called the *row space* of the matrix A.

The column vectors of A are

$$\mathbf{c}_1 = \begin{bmatrix} 3 \\ 0 \end{bmatrix}, \qquad \mathbf{c}_2 = \begin{bmatrix} 1 \\ 1 \end{bmatrix}, \qquad \mathbf{c}_3 = \begin{bmatrix} 1 \\ 1 \end{bmatrix}.$$

The vector space consisting of all linear combinations $\alpha_1\mathbf{c}_1 + \alpha_2\mathbf{c}_2 + \alpha_3\mathbf{c}_3$ is called the *column space* of the matrix A.

Now let V be the vector space spanned by some set of vectors $\{\mathbf{v}_1, \ldots, \mathbf{v}_k\}$, so V consists of all linear combinations of these vectors. It may happen that one of these vectors is itself a linear combination of the others. Suppose for example that

$$\mathbf{v}_1 = \mathbf{v}_2 - 4\mathbf{v}_3.$$

Then we claim that the vector space V can also be spanned by the smaller set of vectors $\{\mathbf{v}_2, \ldots, \mathbf{v}_k\}$. We need only check that every linear combination of the original set of vectors $\{\mathbf{v}_1, \mathbf{v}_2, \ldots, \mathbf{v}_k\}$ can also be expressed as a linear combination of the smaller set $\{\mathbf{v}_2, \ldots, \mathbf{v}_k\}$. But this is obvious, since

$$\begin{aligned} \alpha_1\mathbf{v}_1 + \alpha_2\mathbf{v}_2 + \cdots + \alpha_k\mathbf{v}_k &= \alpha_1(\mathbf{v}_2 - 4\mathbf{v}_3) + \alpha_2\mathbf{v}_2 + \cdots + \alpha_k\mathbf{v}_k \\ &= (\alpha_1 + \alpha_2)\mathbf{v}_2 + (-4\alpha_1 + \alpha_3)\mathbf{v}_3 + \cdots + \alpha_k\mathbf{v}_k. \end{aligned}$$

The same argument works in more general circumstances; it proves that if V is spanned by some set of vectors $\{\mathbf{v}_1, \mathbf{v}_2, \ldots, \mathbf{v}_k\}$, and if one of these vectors is expressible as a linear combination of the others, then that one can be dropped from the spanning set without affecting the vector space V. We can then look at the new spanning set to see whether another vector can be omitted, and so on. This procedure can be continued until no more vectors can be omitted. The process ends when we reach a collection of vectors, chosen from among $\mathbf{v}_1, \ldots, \mathbf{v}_k$, which still span V, but such that no one of them is expressible as a linear combination of the others. We call this a *linearly independent* collection of vectors, and we shall say that this collection is a *basis* for the vector space V.

EXAMPLE

Let

$$A = \begin{bmatrix} 2 & 0 & 1 & -1 \\ 1 & 0 & 1 & -1 \\ 1 & 0 & 0 & 0 \\ 0 & 0 & 1 & -1 \end{bmatrix}.$$

The row vectors of A are

$$\mathbf{r}_1 = (2 \quad 0 \quad 1 \quad -1), \qquad \mathbf{r}_2 = (1 \quad 0 \quad 1 \quad -1),$$
$$\mathbf{r}_3 = (1 \quad 0 \quad 0 \quad 0), \qquad \mathbf{r}_4 = (0 \quad 0 \quad 1 \quad -1).$$

The vector space V spanned by these row vectors is by definition the *row space* of the matrix A. Let us find a basis for V.

We observe first of all that $\mathbf{r}_3 = \mathbf{r}_1 - \mathbf{r}_2$. Therefore we may omit $\mathbf{r}_3$ from the original spanning set $\{\mathbf{r}_1, \mathbf{r}_2, \mathbf{r}_3, \mathbf{r}_4\}$, so as to obtain a smaller set of vectors $\{\mathbf{r}_1, \mathbf{r}_2, \mathbf{r}_4\}$ which also spans V. But now $\mathbf{r}_4 = 2\mathbf{r}_2 - \mathbf{r}_1$, so we can omit $\mathbf{r}_4$ and get a new spanning set $\{\mathbf{r}_1, \mathbf{r}_2\}$. Finally, neither one of the vectors $\mathbf{r}_1$, $\mathbf{r}_2$ is a scalar multiple of the other. Therefore the process terminates, the set $\{\mathbf{r}_1, \mathbf{r}_2\}$ is linearly independent, and is a basis for the row space V of A.

Let us remark that every vector $\mathbf{r}$ in V is expressible as a linear combination of $\mathbf{r}_1$ and $\mathbf{r}_2$, thus:

$$\begin{aligned} \mathbf{r} = \alpha\mathbf{r}_1 + \beta\mathbf{r}_2 &= \alpha(2 \quad 0 \quad 1 \quad -1) + \beta(1 \quad 0 \quad 1 \quad -1) \\ &= (2\alpha + \beta, 0, \alpha + \beta, -\alpha - \beta), \end{aligned}$$

where α, β are scalars. Hence V consists of all vectors of the form

$$(2\alpha + \beta, 0, \alpha + \beta, -\alpha - \beta),$$

where α, β are arbitrary scalars.

It should be pointed out that there are other possible bases for the vector space V. For example, $\{\mathbf{r}_1, \mathbf{r}_4\}$ is a basis for V; so also is $\{\mathbf{r}_2, \mathbf{r}_3\}$.

Let us next find a basis for the column space W of the matrix A. By definition, W consists of all linear combinations of the column vectors $\mathbf{c}_1$, $\mathbf{c}_2$, $\mathbf{c}_3$, $\mathbf{c}_4$ of A. Thus W is spanned by $\{\mathbf{c}_1, \mathbf{c}_2, \mathbf{c}_3, \mathbf{c}_4\}$. We can obviously omit the zero vector $\mathbf{c}_2$ from this spanning set. Furthermore $\mathbf{c}_4 = -\mathbf{c}_3$, so we can also omit $\mathbf{c}_4$. This shows that W is spanned by $\{\mathbf{c}_1, \mathbf{c}_3\}$. Neither of the vectors $\mathbf{c}_1$, $\mathbf{c}_3$ is a scalar multiple of the other, and therefore $\{\mathbf{c}_1, \mathbf{c}_3\}$ is a basis for the column space W of the matrix A.

Now let V be any vector space. The number of vectors in a basis for V is called the *dimension* of the vector space V. This dimension depends only on the vector space V, and not on the procedure used to find of basis of V. We may restate this basic fact as follows:

(9.1) Theorem[1]

If $\{\mathbf{v}_1, \ldots, \mathbf{v}_k\}$ *and* $\{\mathbf{w}_1, \ldots, \mathbf{w}_m\}$ *are two bases for the same vector space* V, *then* $k = m$.

Remarks:

1. The vector space consisting only of the zero vector (in some space) is said to have *dimension zero.*

2. Let V be the vector space consisting of all $1 \times n$ row vectors $\mathbf{v} = (\alpha_1, \ldots, \alpha_n)$. Set

$$\mathbf{e}_1 = (1,0, \ldots ,0), \qquad \mathbf{e}_2 = (0,1,0, \ldots ,0), \ldots ,$$
$$\mathbf{e}_n = (0, \ldots ,0,1).$$

Then $\{\mathbf{e}_1, \mathbf{e}_2, \ldots, \mathbf{e}_n\}$ span V, since we may express each vector $\mathbf{v}$ as a linear combination of the $\mathbf{e}$'s, namely

$$\mathbf{v} = (\alpha_1,\alpha_2, \ldots ,\alpha_n) = \alpha_1\mathbf{e}_1 + \alpha_2\mathbf{e}_2 + \cdots + \alpha_n\mathbf{e}_n.$$

Furthermore, no one of the vectors $\mathbf{e}_1, \ldots, \mathbf{e}_n$ is expressible as a linear combination of the others. Therefore $\{\mathbf{e}_1, \mathbf{e}_2, \ldots, \mathbf{e}_n\}$ is a basis for V, and V is a vector space of dimension n. This shows that our definition of dimension as the number of basis vectors is consistent with our saying that the set of all $1 \times n$ row vectors is an n-dimensional space.

Let us next prove a simple but fundamental property of vector spaces.

(9.1a) Theorem

Let V *be a vector space. Then for any two vectors* $\mathbf{v}, \mathbf{v}'$ *in* V, *each linear combination* $\alpha\mathbf{v} + \beta\mathbf{v}'$ *is also in* V.

*[*PROOF:* According to our definition of vector space, V is the collection of all linear combinations of some set of vectors $\{\mathbf{v}_1, \ldots, \mathbf{v}_k\}$. Since $\mathbf{v}$ and $\mathbf{v}'$ are in V, they may be expressed as

$$\mathbf{v} = \alpha_1\mathbf{v}_1 + \cdots + \alpha_k\mathbf{v}_k, \qquad \mathbf{v}' = \beta_1\mathbf{v}_1 + \cdots + \beta_k\mathbf{v}_k,$$

where $\alpha_1, \ldots, \alpha_k, \beta_1, \ldots, \beta_k$ are scalars. Now let α, β be any scalars. Then

$$\alpha\mathbf{v} + \beta\mathbf{v}' = (\alpha\alpha_1 + \beta\beta_1)\mathbf{v}_1 + \cdots + (\alpha\alpha_k + \beta\beta_k)\mathbf{v}_k.$$

This shows that $\alpha\mathbf{v} + \beta\mathbf{v}'$ is also a linear combination of $\mathbf{v}_1, \ldots, \mathbf{v}_k$, and therefore lies in V. This completes the proof.]*

[1] See references listed in the Preface for proof.

The property of a vector space described in (9.1a) is important because it characterizes vector spaces. Indeed, we have the following fundamental result:

(9.2) Theorem[2]

Let S be some nonempty collection of vectors in an n-dimensional space, having the property that for each pair of vectors $\mathbf{v}$, $\mathbf{v}'$ *in S, every linear combination* $\alpha\mathbf{v} + \beta\mathbf{v}'$ *also lies in S. Then S is a vector space, and has a basis of at most n vectors.*

EXAMPLE

Let $B^{m\times n}$ be a fixed matrix, and let S consist of all $n \times 1$ vectors $\mathbf{v}$ such that $B\mathbf{v} = \mathbf{0}$. We claim that S is a vector space. Indeed, let $\mathbf{v}$ and $\mathbf{v}'$ lie in S, so $B\mathbf{v} = \mathbf{0}$ and $B\mathbf{v}' = \mathbf{0}$. But then

$$B(\alpha\mathbf{v} + \beta\mathbf{v}') = \alpha B\mathbf{v} + \beta B\mathbf{v}' = \mathbf{0},$$

so also $\alpha\mathbf{v} + \beta\mathbf{v}$ lies in S. Further, S is nonempty since $\mathbf{0}$ lies in S. Therefore S is a vector space, by Theorem 9.2.

In some of the earlier examples in this section, we have already seen that we may associate with an $m \times n$ matrix A two vector spaces, as follows: the *row space* of A is the vector space spanned by the row vectors of A, while the *column space* of A is the vector space spanned by the column vectors of A. We have also discussed briefly how we may find bases for these row spaces and column spaces. Let us give a few more examples to illustrate this procedure.

EXAMPLES

1. $A = \begin{bmatrix} 3 & 1 & 1 \\ 0 & 1 & 1 \end{bmatrix}$. The row vectors of A are

$$\mathbf{r}_1 = [3 \quad 1 \quad 1], \qquad \mathbf{r}_2 = [0 \quad 1 \quad 1],$$

and the row space of A consists of all linear combinations

$$\alpha\mathbf{r}_1 + \beta\mathbf{r}_2 = [3\alpha,\ \alpha+\beta,\ \alpha+\beta], \quad \alpha, \beta \text{ scalars.}$$

Since neither row vector is a scalar multiple of the other, the set $\{\mathbf{r}_1, \mathbf{r}_2\}$ is a basis for the row space.

The column vectors of A are

$$\mathbf{c}_1 = \begin{bmatrix} 3 \\ 0 \end{bmatrix}, \quad \mathbf{c}_2 = \begin{bmatrix} 1 \\ 1 \end{bmatrix}, \quad \mathbf{c}_3 = \begin{bmatrix} 1 \\ 1 \end{bmatrix},$$

and the column space of A is the vector space spanned by $\mathbf{c}_1$, $\mathbf{c}_2$, and $\mathbf{c}_3$.

[2] See references in the Preface for proof.

Since $\mathbf{c}_3 = \mathbf{c}_2$, we omit $\mathbf{c}_3$ from this spanning set. Clearly the set $\{\mathbf{c}_1, \mathbf{c}_2\}$ is linearly independent, and thus, is a basis for the column space of A.

2.
$$A = \begin{bmatrix} 1 & 0 & 0 & 1 \\ 2 & 0 & 1 & 2 \\ 0 & 0 & 1 & 0 \end{bmatrix},$$

$$\mathbf{c}_1 = \begin{bmatrix} 1 \\ 2 \\ 0 \end{bmatrix}, \quad \mathbf{c}_2 = \begin{bmatrix} 0 \\ 0 \\ 0 \end{bmatrix}, \quad \mathbf{c}_3 = \begin{bmatrix} 0 \\ 1 \\ 1 \end{bmatrix}, \quad \mathbf{c}_4 = \begin{bmatrix} 1 \\ 2 \\ 0 \end{bmatrix},$$

$$\mathbf{r}_1 = [1 \quad 0 \quad 0 \quad 1], \qquad \mathbf{r}_2 = [2 \quad 0 \quad 1 \quad 2], \qquad \mathbf{r}_3 = [0 \quad 0 \quad 1 \quad 0].$$

Note that $\mathbf{r}_2 = 2\mathbf{r}_1 + \mathbf{r}_3$, so that the row space of A has basis $\{\mathbf{r}_1, \mathbf{r}_3\}$. On the other hand, $\mathbf{c}_4 = \mathbf{c}_1$ and $\mathbf{c}_2 = \mathbf{0}$, so the column space of A has basis $\{\mathbf{c}_1, \mathbf{c}_3\}$.

In the above examples, the row space of A has the same dimension as the column space of A. This is true in general:

(9.3) Theorem[3]

The dimension of the row space of any matrix equals the dimension of its column space.

The *rank* of a matrix is the dimension of its row space, or equivalently, the dimension of its column space. If A has rank r, then the vector space spanned by its row vectors has dimension r. This means that we can find r rows of A such that each of the other rows of A is expressible as a linear combination of these r rows. However, we cannot find a set of $r - 1$ rows with this property, for if we could do so, the dimension of the row space of A would be at most $r - 1$.

Let us show that if we add to one row of a matrix A some multiple of another row, the matrix we obtain will have the same row space as A, and thus will also have the same rank as A. For simplicity, let A have three rows $\mathbf{r}_1$, $\mathbf{r}_2$, $\mathbf{r}_3$, and let B be the matrix with rows $\mathbf{r}_1 + c\mathbf{r}_2$, $\mathbf{r}_2$, $\mathbf{r}_3$. In order to show that A and B have the same row space, we must prove that the set of vectors $\{\mathbf{r}_1, \mathbf{r}_2, \mathbf{r}_3\}$ spans the same vector space as does the set $\{\mathbf{r}_1 + c\mathbf{r}_2, \mathbf{r}_2, \mathbf{r}_3\}$. We may prove this by showing that every linear combination of the first set of vectors is also a linear combination of the second set of vectors, and conversely. We have

$$\alpha_1\mathbf{r}_1 + \alpha_2\mathbf{r}_2 + \alpha_3\mathbf{r}_3 = \alpha_1(\mathbf{r}_1 + c\mathbf{r}_2) + (\alpha_2 - c\alpha_1)\mathbf{r}_2 + \alpha_3\mathbf{r}_3,$$

$$\beta_1(\mathbf{r}_1 + c\mathbf{r}_2) + \beta_2\mathbf{r}_2 + \beta_3\mathbf{r}_3 = \beta_1\mathbf{r}_1 + (c\beta_1 + \beta_2)\mathbf{r}_2 + \beta_3\mathbf{r}_3.$$

The first equation proves that every linear combination of $\{\mathbf{r}_1, \mathbf{r}_2, \mathbf{r}_3\}$

[3] See references in the Preface for proofs.

is also expressible as a linear combination of $\{\mathbf{r}_1 + c\mathbf{r}_2, \mathbf{r}_2, \mathbf{r}_3\}$, while the second equation proves the converse.

We have now proved that adding multiples of one row to another row does not affect the row space of a matrix. We may therefore use our method of "elimination of variables" as a systematic way of finding a basis for the row space of a matrix A. We need only perform the same operations on the rows of A that we would use in eliminating variables in a system of simultaneous linear equations having A as matrix of coefficients.

EXAMPLE

$$A = \begin{bmatrix} 1 & 0 & 2 & 3 & 1 \\ 2 & 0 & 4 & 7 & 3 \\ 0 & 0 & 0 & 1 & 1 \\ 0 & 0 & 0 & 0 & 1 \\ 0 & 0 & 0 & 3 & 4 \\ 1 & 0 & 2 & 1 & 1 \end{bmatrix} \xrightarrow[r_2 - 2r_1]{} \begin{bmatrix} 1 & 0 & 2 & 3 & 1 \\ 0 & 0 & 0 & 1 & 1 \\ 0 & 0 & 0 & 1 & 1 \\ 0 & 0 & 0 & 0 & 1 \\ 0 & 0 & 0 & 3 & 4 \\ 1 & 0 & 2 & 1 & 1 \end{bmatrix}$$

$$\xrightarrow[r_6 - r_1]{} \begin{bmatrix} 1 & 0 & 2 & 3 & 1 \\ 0 & 0 & 0 & 1 & 1 \\ 0 & 0 & 0 & 1 & 1 \\ 0 & 0 & 0 & 0 & 1 \\ 0 & 0 & 0 & 3 & 4 \\ 0 & 0 & 0 & -2 & 0 \end{bmatrix} \xrightarrow[r_3 - r_2]{} \begin{bmatrix} 1 & 0 & 2 & 3 & 1 \\ 0 & 0 & 0 & 1 & 1 \\ 0 & 0 & 0 & 0 & 0 \\ 0 & 0 & 0 & 0 & 1 \\ 0 & 0 & 0 & 3 & 4 \\ 0 & 0 & 0 & -2 & 0 \end{bmatrix}$$

$$\xrightarrow[r_6 + 2r_2]{r_5 - 3r_2} \begin{bmatrix} 1 & 0 & 2 & 3 & 1 \\ 0 & 0 & 0 & 1 & 1 \\ 0 & 0 & 0 & 0 & 0 \\ 0 & 0 & 0 & 0 & 1 \\ 0 & 0 & 0 & 0 & 1 \\ 0 & 0 & 0 & 0 & 2 \end{bmatrix} \xrightarrow[r_6 - 2r_4]{r_5 - r_4} \begin{bmatrix} 1 & 0 & 2 & 3 & 1 \\ 0 & 0 & 0 & 1 & 1 \\ 0 & 0 & 0 & 0 & 0 \\ 0 & 0 & 0 & 0 & 1 \\ 0 & 0 & 0 & 0 & 0 \\ 0 & 0 & 0 & 0 & 0 \end{bmatrix}.$$

Therefore the row space of A has basis

$$[1 \;\; 0 \;\; 2 \;\; 3 \;\; 1], [0 \;\; 0 \;\; 0 \;\; 1 \;\; 1], [0 \;\; 0 \;\; 0 \;\; 0 \;\; 1],$$

and A has rank 3.

Let us give another description of the rank of an $m \times n$ matrix A in terms of the minor determinants of A. By a $k \times k$ *minor* of A we mean a determinant whose entries lie at the positions where k rows of A cross k columns of A. For example, let

$$A = \begin{bmatrix} a_{11} & a_{12} & a_{13} & a_{14} & a_{15} \\ a_{21} & a_{22} & a_{23} & a_{24} & a_{25} \\ a_{31} & a_{32} & a_{33} & a_{34} & a_{35} \\ a_{41} & a_{42} & a_{43} & a_{44} & a_{45} \end{bmatrix} \begin{matrix} \\ \leftarrow \\ \\ \leftarrow \end{matrix}$$

(arrows ↓ ↓ mark columns 2 and 3; rows 2 and 4 and columns 2 and 3 are shaded)

Then $\begin{vmatrix} a_{22} & a_{23} \\ a_{42} & a_{43} \end{vmatrix}$ is a 2×2 minor of A. Some of the other 2×2 minors are

$$\begin{vmatrix} a_{11} & a_{14} \\ a_{31} & a_{34} \end{vmatrix}, \begin{vmatrix} a_{12} & a_{15} \\ a_{22} & a_{25} \end{vmatrix}, \begin{vmatrix} a_{32} & a_{33} \\ a_{42} & a_{43} \end{vmatrix}.$$

(9.4) Theorem[4]

Let $A^{m \times n}$ be a matrix of rank r. Then

(i) $r \leqslant m$ *and* $r \leqslant n$;
(ii) *the row space of A has dimension r; in other words, there exist r rows of A such that each of the other rows is a linear combination of these r rows, but no smaller number of rows could be used for this purpose;*
(iii) *the column space of A has dimension r;*
(iv) *every* $(r+1) \times (r+1)$ *minor of A is* 0, *but at least one* $r \times r$ *minor is not* 0;
(v) *an* $n \times n$ *matrix is nonsingular if and only if it has rank n.*

Our definition of the rank of a matrix A (as the dimension of its row space) agrees with the definition given in the discussion preceding Theorem 7.13. It is therefore not surprising that the ideas in this section are closely related to the basic problem of solving a system of m simultaneous linear equations in n unknowns $x_1, \ldots, x_n$:

$$\text{(9.5)} \quad \begin{cases} a_{11}x_1 + a_{12}x_2 + \cdots + a_{1n}x_n = b_1 \\ a_{21}x_1 + a_{22}x_2 + \cdots + a_{2n}x_n = b_2 \\ \qquad \cdots \\ a_{m1}x_1 + a_{m2}x_2 + \cdots + a_{mn}x_n = b_m. \end{cases}$$

Let $A = [a_{ij}]^{m \times n}$ be the matrix of coefficients of the unknowns. Define the *augmented matrix* of this system of equations to be

$$[A \quad \mathbf{b}] = \begin{bmatrix} a_{11} & a_{12} & \cdots & a_{1n} & b_1 \\ a_{21} & a_{22} & \cdots & a_{2n} & b_2 \\ & & \cdots & & \\ a_{m1} & a_{m2} & \cdots & a_{mn} & b_m \end{bmatrix}.$$

(9.6) Theorem[5]

The system (9.5) *of simultaneous linear equations is consistent (and has a solution) if and only if the augmented matrix* $[A \quad \mathbf{b}]$ *has the same rank as the matrix of coefficients A.*

[4] See references in the Preface. Part (ii) just repeats the definition of rank, and is included here for the convenience of the reader. We have already proved part (v) in (7.13) and (7.15).

[5] See references in the Preface for proofs of this theorem and Theorem (9.7).

(9.7) Theorem

In the system (9.5), *suppose that* $[A \quad \mathbf{b}]$ *and* A *both have rank* r. *Then the system is consistent, and we can solve for precisely* r *of the unknowns uniquely in terms of the remaining* $n-r$ *unknowns, which may have arbitrary values.*

Let us illustrate the above theorems. Consider first the simultaneous equations

$$\textbf{(9.8)} \qquad \begin{cases} x + 2y = 3 \\ 2x + 4y = 7, \end{cases}$$

with matrix of coefficients $A = \begin{bmatrix} 1 & 2 \\ 2 & 4 \end{bmatrix}$, and augmented matrix $\begin{bmatrix} 1 & 2 & 3 \\ 2 & 4 & 7 \end{bmatrix}$. Then A has rank 1, while the augmented matrix has rank 2. Thus by Theorem 9.6, the system (9.8) is inconsistent. This is obvious, in any case, since if there were a solution, then subtracting twice the first equation from the second would give the contradiction $0 = 1$.

On the other hand, the system of simultaneous equations

$$\textbf{(9.9)} \qquad \begin{cases} x + 2y = 3 \\ 2x + 4y = 6 \end{cases}$$

has the same matrix of coefficients A as above, of rank 1. The augmented matrix is $\begin{bmatrix} 1 & 2 & 3 \\ 2 & 4 & 6 \end{bmatrix}$, also of rank 1. Thus by the above theorems, the system (9.9) is consistent, and we can solve for 1 of the variables ($r = 1$) in terms of the remaining variable ($n - r = 1$). Indeed, on subtracting the first equation from the second equation, the new second equation is the identity $0 = 0$. Thus the solution of (9.9) is given by $x = 3 - 2y$, y arbitrary. We could also express the solution as $y = \frac{1}{2}(3 - x)$, x arbitrary.

*[We shall conclude this section with some examples of vector spaces which arise naturally in calculus. Such vector spaces are usually of infinite dimension, and so we can no longer stay within the framework of an n-dimensional space. We shall therefore extend our definition of a vector space, guided by Theorem 9.2. Let us agree to *define* a vector space as a nonempty collection V of objects (= "vectors"), such that for each pair of objects $\mathbf{v}$, $\mathbf{v}'$ in V, every linear combination $\alpha\mathbf{v} + \beta\mathbf{v}'$ also lies in V.

In the following discussion, the scalars α, β, and so on, are assumed to be real numbers, for convenience. By a *real polynomial* we mean an expression of the form $\alpha_0 + \alpha_1 x + \cdots + \alpha_n x^n$ in which each α_j is real.

* The remainder of this section is optional material.

(9.10) EXAMPLE

Let V be the set of all real polynomials in a variable x. Then V is a vector space, since if $f(x)$ and $g(x)$ are real polynomials, then so is each linear combination $\alpha f(x) + \beta g(x)$.

(9.11) EXAMPLE

Let W be the set of all real polynomials of degree at most 2. Each "vector" $f(x)$ in W is uniquely expressible in the form

$$f(x) = \alpha_0 \cdot 1 + \alpha_1 \cdot x + \alpha_2 \cdot x^2, \quad \alpha_0, \alpha_1, \alpha_2 \text{ real}.$$

This means that $\{1, x, x^2\}$ form a basis for W, and W is a vector space of dimension 3.

(9.12) EXAMPLE

Let R be the set of all real numbers. Then R is a vector space of dimension one, with basis $\{1\}$.

(9.13) EXAMPLE

Let $C[0,1]$ be the set of all real-valued functions defined and continuous at every point of the closed interval $[0,1]$. Since linear combinations of continuous functions are also continuous, it is clear that $C[0,1]$ is a vector space.

(9.14) EXAMPLE

Let S be the set of all functions f, continuous at every point of the closed interval $[0,1]$, which satisfy the condition that

$$\int_0^1 f(x)\, dx = 0.$$

If f and g are in S, and α, β are scalars, then

$$\int_0^1 (\alpha f(x) + \beta g(x))\, dx = \alpha \int_0^1 f(x)\, dx + \beta \int_0^1 g(x)\, dx = 0.$$

This shows that $\alpha f + \beta g$ is also in S, so S is a vector space.]*

EXERCISES

1. Find a basis for the vector space spanned by

$$\mathbf{v}_1 = (2 \quad 1 \quad 1 \quad -1), \qquad \mathbf{v}_2 = (2 \quad 1 \quad 0 \quad 0), \qquad \mathbf{v}_3 = (0 \quad 0 \quad 1 \quad -1).$$

2. Find bases for the row spaces, and also for the column spaces, of each of the following matrices:

$$[0 \quad 1], \; [1 \quad 1], \; \begin{bmatrix} 2 & 1 \\ 1 & 2 \end{bmatrix}, \; \begin{bmatrix} 1 & 0 & 0 & 1 \\ 0 & 0 & 0 & 1 \\ 0 & 1 & 0 & 1 \end{bmatrix}, \; \begin{matrix} 1 & 0 & 1 \\ 0 & 1 & 1 \\ 1 & 0 & -1 \\ 0 & 1 & -1 \end{matrix}, \; \begin{bmatrix} 0 & 1 & 0 \\ 0 & 3 & 0 \end{bmatrix}.$$

Give the rank of each matrix.

3. For each of the above matrices, find a nonzero minor of largest possible size. ("Size" refers to the number of rows and columns.)
4. If A and B are $m \times n$ matrices, each of rank r, what can be said about the rank of $A + B$? of $2A$?
5. Show that if C is a submatrix of A, then rank of $C \leqslant$ rank of A.

*6. What is the rank of A, where

$$A = \begin{bmatrix} 0 & 1 & 0 & \cdots & 0 & 0 \\ 0 & 0 & 1 & \cdots & 0 & 0 \\ & & & \cdots & & \\ 0 & 0 & 0 & \cdots & 0 & 1 \\ 0 & 0 & 0 & \cdots & 0 & 0 \end{bmatrix}^{n \times n} ?$$

Determine the matrices $A^2, A^3, \ldots, A^n$ and their ranks.

7. Find two square matrices A, B, each of rank 1, such that $AB = 0$.
8. Consider the system of simultaneous linear equations

$$a_1x + b_1y + c_1z = 0, \qquad a_2x + b_2y + c_2z = 0, \qquad a_3x + b_3y + c_3z = 0.$$

Solving them simultaneously amounts to finding the points of intersection of three planes through the origin. What is the geometric significance of the rank of the matrix of coefficients?

9. Give an analogous geometrical discussion for a system of three nonhomogeneous equations in three unknowns.
10. Let $\mathbf{r}_1, \ldots, \mathbf{r}_k$ be a set of $1 \times n$ row vectors. Show that they form a linearly independent set of vectors if and only if the $k \times n$ matrix with rows $\mathbf{r}_1, \ldots, \mathbf{r}_k$ has rank k.
11. Show that $n + 1$ vectors, each of size $1 \times n$, cannot form a linearly independent set of vectors.
12. Let $\mathbf{r}_1, \ldots, \mathbf{r}_n$ be the rows of an $n \times n$ matrix A. Prove that A is singular if and only if there exist constants $c_1, \ldots, c_n$, not all zero, such that $c_1\mathbf{r}_1 + \cdots + c_n\mathbf{r}_n = \mathbf{0}$.

13 Let $\{\mathbf{v}_1, \ldots, \mathbf{v}_k\}$ be a basis for the vector space V. Show that if $\alpha_1\mathbf{v}_1 + \cdots + \alpha_k\mathbf{v}_k = \beta_1\mathbf{v}_1 + \cdots + \beta_k\mathbf{v}_k$, then $\alpha_1 = \beta_1, \ldots, \alpha_k = \beta_k$. Deduce from this that every vector in V is expressible as a linear combination of the basis vectors in exactly one way.

*14. Let $\{\mathbf{v}_1, \ldots, \mathbf{v}_k\}$ be a basis for the vector space V, and let $c_2, \ldots, c_k$ be scalars. Show that

$$\{\mathbf{v}_1 + c_2\mathbf{v}_2 + \cdots + c_k\mathbf{v}_k, \mathbf{v}_2, \ldots, \mathbf{v}_k\}$$

is also a basis for V.

*15. Show that the rank of the matrix

$$\begin{bmatrix} A^{m\times m} & 0^{m\times n} \\ 0^{n\times m} & B^{n\times n} \end{bmatrix}$$

is equal to the rank of A plus the rank of B.

*16. Show that adjoining to a matrix one or more columns of zeros does not change the rank of the matrix.

*17. Let $A^{m\times n}$ be a matrix of rank r, and consider the system of simultaneous equations (in the unknowns $x_1, \ldots, x_n$)

$$A \cdot \begin{bmatrix} x_1 \\ \cdot \\ \cdot \\ \cdot \\ x_n \end{bmatrix} = \begin{bmatrix} 0 \\ \cdot \\ \cdot \\ \cdot \\ 0 \end{bmatrix}.$$

Show that the system is consistent, and that we can solve for uniquely r of the unknowns $x_1, \ldots, x_n$ in terms of the remaining $n - r$ unknowns.

*18. Let $A^{m\times n}$ be a matrix of rank r. Show that its transpose A^{T} also has rank r.

*19. Let $A^{m\times n}$ have rank r. Discuss the solution of the system of equations (in the unknowns $y_1, \ldots, y_m$) given by

$$[y_1 \quad \cdots \quad y_m]A = [0 \quad \cdots \quad 0].$$

*20. Show that each of the following is a vector space:

(a) The set of all real polynomials of degree at most 4.

(b) The set of all functions f, defined at every point of the closed interval $[0,1]$, which satisfy the condition that $f(\frac{1}{2}) = 0$.

(c) The set of all functions f, defined at every point of the closed interval $[0,1]$, satisfying the condition that $f(0) = f(1)$.

(d) The set of all functions f, defined and differentiable at every point of the closed interval $[0,1]$.

(e) The set of all power series $\sum_{n=0}^{\infty} a_n x^n$ which converge for $|x| \leqslant 1$.

(f) The set of all odd functions. (Call f *odd* if it satisfies the identity $f(-x) = -f(x)$ for all x.)

(g) The set of all even functions. (Call f *even* if $f(-x) = f(x)$ for all x.)

*21. Show that the following are *not* vector spaces:

(a) The set of all real polynomials of degree exactly 2.

(b) The set of all functions f, defined at every point of the closed interval $[0,1]$, satisfying the condition that $f(\frac{1}{2}) > 0$.

(c) The set of all functions f for which $f(0) = 1$.

10 KERNEL AND RANGE OF A LINEAR TRANSFORMATION

Let A be a given $m \times n$ matrix, and let $\mathbf{v}$ denote an arbitrary $n \times 1$ column vector. Then $A\mathbf{v}$ is an $m \times 1$ column vector. The mapping

$$\mathbf{v} \longrightarrow A\mathbf{v}, \tag{10.1}$$

which maps $\mathbf{v}$ onto $A\mathbf{v}$, is called a *linear transformation* from n-space into m-space. We have already encountered this concept in Section 8 for the special case where $m = n = 2$.

EXAMPLE

$$A = \begin{bmatrix} 1 & -1 \\ 2 & 0 \\ 1 & 2 \end{bmatrix}, \qquad \mathbf{v} = \begin{bmatrix} x \\ y \end{bmatrix}, \qquad A\mathbf{v} = \begin{bmatrix} 1 & -1 \\ 2 & 0 \\ 1 & 2 \end{bmatrix} \begin{bmatrix} x \\ y \end{bmatrix} = \begin{bmatrix} x - y \\ 2x \\ x + 2y \end{bmatrix}.$$

Thus A maps $\begin{bmatrix} x \\ y \end{bmatrix}$ onto $\begin{bmatrix} x - y \\ 2x \\ x + 2y \end{bmatrix}$.

Returning to the general case of the linear transformation (10.1), let us observe that if $\mathbf{v}$, $\mathbf{v}'$ are any $n \times 1$ column vectors, and α, β are scalars, then

$$A(\alpha\mathbf{v} + \beta\mathbf{v}') = A(\alpha\mathbf{v}) + A(\beta\mathbf{v}') = \alpha(A\mathbf{v}) + \beta(A\mathbf{v}').$$

In other words, A maps any linear combination of the vectors $\mathbf{v}$ and $\mathbf{v}'$ onto the corresponding linear combination of $A\mathbf{v}$ and $A\mathbf{v}'$. We are now going to show that this fundamental property characterizes linear transformations.

(10.2) Theorem

Let F be a function which maps each vector $\mathbf{v}$ in n-space onto a vector $F(\mathbf{v})$ in m-space. Suppose that F satisfies the identity

$$F(\alpha\mathbf{v} + \beta\mathbf{v}') = \alpha F(\mathbf{v}) + \beta F(\mathbf{v}') \tag{10.3}$$

for all vectors $\mathbf{v}$, $\mathbf{v}'$ *in n-space, and all scalars* α, β. *Then F is a linear transformation, that is, there exists a matrix* $B^{m\times n}$ *such that*

(10.4) $$F(\mathbf{v}) = B\mathbf{v} \quad \text{for all } \mathbf{v}.$$

PROOF: For simplicity of notation, consider the case where $m = n = 2$. Let

$$\mathbf{v}_1 = \begin{bmatrix} 1 \\ 0 \end{bmatrix}, \quad \mathbf{v}_2 = \begin{bmatrix} 0 \\ 1 \end{bmatrix}, \quad F(\mathbf{v}_1) = \begin{bmatrix} p \\ r \end{bmatrix}, \quad F(\mathbf{v}_2) = \begin{bmatrix} q \\ s \end{bmatrix},$$

and set $B = \begin{bmatrix} p & q \\ r & s \end{bmatrix}$. We shall show that (10.4) holds true with this choice of B. Let $\mathbf{v}$ be any vector in 2-space. Then we may write

$$\mathbf{v} = \begin{bmatrix} \alpha \\ \beta \end{bmatrix} = \alpha\mathbf{v}_1 + \beta\mathbf{v}_2,$$

so by (10.3), we have

$$F(\mathbf{v}) = F(\alpha\mathbf{v}_1 + \beta\mathbf{v}_2) = \alpha F(\mathbf{v}_1) + \beta F(\mathbf{v}_2) = \alpha \begin{bmatrix} p \\ r \end{bmatrix} + \beta \begin{bmatrix} q \\ s \end{bmatrix}$$
$$= \begin{bmatrix} p & q \\ r & s \end{bmatrix} \begin{bmatrix} \alpha \\ \beta \end{bmatrix} = B\mathbf{v}.$$

This establishes the theorem for the case $m = n = 2$. The proof for general m,n is similar.

Many authors *define* a linear transformation as a function F satisfying the identity (10.3). We have just shown that, for finite dimensional vector spaces, this is equivalent to our original definition (10.1).

EXAMPLES

1. Find a linear transformation from 3-space to 2-space such that

$$\begin{bmatrix} 1 \\ 0 \\ 0 \end{bmatrix} \longrightarrow \begin{bmatrix} 2 \\ 3 \end{bmatrix}, \quad \begin{bmatrix} 0 \\ 1 \\ 0 \end{bmatrix} \longrightarrow \begin{bmatrix} -1 \\ 2 \end{bmatrix}, \quad \begin{bmatrix} 0 \\ 0 \\ 1 \end{bmatrix} \longrightarrow \begin{bmatrix} 4 \\ 1 \end{bmatrix}.$$

We must find a matrix $A^{2\times 3}$ for which

$$A \cdot \begin{bmatrix} 1 \\ 0 \\ 0 \end{bmatrix} = \begin{bmatrix} 2 \\ 3 \end{bmatrix}, \quad A \cdot \begin{bmatrix} 0 \\ 1 \\ 0 \end{bmatrix} = \begin{bmatrix} -1 \\ 2 \end{bmatrix}, \quad A \cdot \begin{bmatrix} 0 \\ 0 \\ 1 \end{bmatrix} = \begin{bmatrix} 4 \\ 1 \end{bmatrix}.$$

Together, these three conditions can be written as a single matrix equation

$$A \cdot \begin{bmatrix} 1 & 0 & 0 \\ 0 & 1 & 0 \\ 0 & 0 & 1 \end{bmatrix} = \begin{bmatrix} 2 & -1 & 4 \\ 3 & 2 & 1 \end{bmatrix}.$$

But the left-hand expression equals A, so $A = \begin{bmatrix} 2 & -1 & 4 \\ 3 & 2 & 1 \end{bmatrix}$, and the linear transformation is given by

$$\begin{bmatrix} x \\ y \\ z \end{bmatrix} \longrightarrow \begin{bmatrix} 2 & -1 & 4 \\ 3 & 2 & 1 \end{bmatrix} \begin{bmatrix} x \\ y \\ z \end{bmatrix} = \begin{bmatrix} 2x - y + 4z \\ 3x + 2y + z \end{bmatrix}.$$

2. Find a matrix $A^{4\times 3}$ satisfying

$$A \cdot \begin{bmatrix} 1 \\ 1 \\ 0 \end{bmatrix} = \begin{bmatrix} 1 \\ 0 \\ 1 \\ 1 \end{bmatrix}, \quad A \cdot \begin{bmatrix} 0 \\ 1 \\ 3 \end{bmatrix} = \begin{bmatrix} 3 \\ 1 \\ 0 \\ 0 \end{bmatrix}, \quad A \cdot \begin{bmatrix} 0 \\ 0 \\ 1 \end{bmatrix} = \begin{bmatrix} 0 \\ 0 \\ 4 \\ 1 \end{bmatrix}.$$

These give

$$A \cdot \begin{bmatrix} 1 & 0 & 0 \\ 1 & 1 & 0 \\ 0 & 3 & 1 \end{bmatrix} = \begin{bmatrix} 1 & 3 & 0 \\ 0 & 1 & 0 \\ 1 & 0 & 4 \\ 1 & 0 & 1 \end{bmatrix}, \qquad A = \begin{bmatrix} 1 & 3 & 0 \\ 0 & 1 & 0 \\ 1 & 0 & 4 \\ 1 & 0 & 1 \end{bmatrix} \begin{bmatrix} 1 & 0 & 0 \\ 1 & 1 & 0 \\ 0 & 3 & 1 \end{bmatrix}^{-1}.$$

Routine calculation yields

$$\begin{bmatrix} 1 & 0 & 0 \\ 1 & 1 & 0 \\ 0 & 3 & 1 \end{bmatrix}^{-1} = \begin{bmatrix} 1 & 0 & 0 \\ -1 & 1 & 0 \\ 3 & -3 & 1 \end{bmatrix}, \qquad A = \begin{bmatrix} -2 & 3 & 0 \\ -1 & 1 & 0 \\ 13 & -12 & 4 \\ 4 & -3 & 1 \end{bmatrix}.$$

The *range* of the linear transformation (10.1) is the set of all vectors of the form $A\mathbf{v}$ for some $\mathbf{v}$ in n-space.

(10.5) Theorem

The range of the linear transformation (10.1) *is the column space*[1] *of the matrix A. The dimension of the range equals the rank of A.*

*[*PROOF:* A numerical example will convince us of the truth of the first part of the theorem. Let $A = \begin{bmatrix} 2 & -1 & 4 \\ 1 & 3 & 1 \end{bmatrix}$, and let $\mathbf{v} = \begin{bmatrix} x \\ y \\ z \end{bmatrix}$ be a vector in 3-space. The linear transformation determined by A maps $\mathbf{v}$ onto the vector $A\mathbf{v}$ in 2-space. By definition, the range of the linear transformation consists of all vectors of the form $A\mathbf{v}$, where $\mathbf{v}$ is an arbitrary vector in 3-space.

[1] Recall that the *column space* of A is the vector space spanned by the column vectors of A. The dimension of this column space equals the rank of A, by (9.4).

Now

$$A\mathbf{v} = \begin{bmatrix} 2 & -1 & 4 \\ 1 & 3 & 1 \end{bmatrix} \begin{bmatrix} x \\ y \\ z \end{bmatrix} = \begin{bmatrix} 2x - y + 4z \\ x + 3y + z \end{bmatrix}$$

$$= x \begin{bmatrix} 2 \\ 1 \end{bmatrix} + y \begin{bmatrix} -1 \\ 3 \end{bmatrix} + z \begin{bmatrix} 4 \\ 1 \end{bmatrix}.$$

Thus the range consists of all linear combinations of the vectors $\begin{bmatrix} 2 \\ 1 \end{bmatrix}$, $\begin{bmatrix} -1 \\ 3 \end{bmatrix}$, $\begin{bmatrix} 4 \\ 1 \end{bmatrix}$. But these are the column vectors of A, and the vector space they span is by definition the column space of A. This shows that the range equals the column space of A. A similar argument works in the general case.

The second statement in the theorem is a direct consequence of Theorem 9.4, which states that the dimension of the column space of A equals the rank of A.]*

Let $A^{m \times n}$ be an arbitrary matrix, and let $\mathbf{v}$ denote a vector in n-space. The linear transformation determined by A maps each vector $\mathbf{v}$ onto the vector $A\mathbf{v}$ in m-space. We define the *kernel* of the linear transformation as the set of all vectors $\mathbf{v}$ for which $A\mathbf{v} = \mathbf{0}$. This kernel is also called the *null space* of the matrix A.

(10.5a) EXAMPLE

Let $A = \begin{bmatrix} 2 & 1 & 1 & 1 \\ 1 & 0 & 1 & -1 \end{bmatrix}^{2 \times 4}$. To find the null space of A, we must determine all vectors $\mathbf{v}$ in 4-space for which $A\mathbf{v} = \mathbf{0}$. We may write this condition as

$$A\mathbf{v} = \begin{bmatrix} 2 & 1 & 1 & 1 \\ 1 & 0 & 1 & -1 \end{bmatrix} \begin{bmatrix} x \\ y \\ z \\ w \end{bmatrix} = \begin{bmatrix} 2x + y + z + w \\ x + z - w \end{bmatrix} = \begin{bmatrix} 0 \\ 0 \end{bmatrix},$$

that is,

$$\begin{cases} 2x + y + z + w = 0 \\ x \quad\quad + z - w = 0. \end{cases}$$

These simultaneous equations have the solution

$$x = -z + w, \qquad y = z - 3w, \qquad z, w \text{ arbitrary.}$$

Therefore the null space of A consists of all vectors

$$\mathbf{v} = \begin{bmatrix} -z+w \\ z-3w \\ z \\ w \end{bmatrix} = z\begin{bmatrix} -1 \\ 1 \\ 1 \\ 0 \end{bmatrix} + w\begin{bmatrix} 1 \\ -3 \\ 0 \\ 1 \end{bmatrix},$$

where z, w are arbitrary. Let us write

$$\mathbf{v}_1 = \begin{bmatrix} -1 \\ 1 \\ 1 \\ 0 \end{bmatrix}, \qquad \mathbf{v}_2 = \begin{bmatrix} 1 \\ -3 \\ 0 \\ 1 \end{bmatrix}.$$

Then the null space of A is the vector space spanned by $\{\mathbf{v}_1, \mathbf{v}_2\}$. Neither of $\mathbf{v}_1$, $\mathbf{v}_2$ is a scalar multiple of the other, so the set $\{\mathbf{v}_1, \mathbf{v}_2\}$ is linearly independent and is a basis for the null space of A. Since there are 2 vectors in this basis, the null space has dimension 2.

We now prove a basic theorem which relates the rank of the matrix A with the dimension of the null space of A.

(10.6) Theorem

Let $A^{m\times n}$ be a matrix of rank r. Then the null space of A is a vector space of dimension $n-r$. In other words, the kernel of the linear transformation $\mathbf{v} \to A\mathbf{v}$ is a vector space of dimension $n-r$.

*[*PROOF:* The null space of A equals the kernel of the linear transformation $\mathbf{v} \to A\mathbf{v}$, and consists of all vectors $\mathbf{v}$ in n-space such that $A\mathbf{v} = \mathbf{0}$. Let us write

$$A\mathbf{v} = \begin{bmatrix} a_{11} & \cdots & a_{1n} \\ & \cdots & \\ a_{m1} & \cdots & a_{mn} \end{bmatrix} \begin{bmatrix} x_1 \\ \cdot \\ \cdot \\ \cdot \\ x_n \end{bmatrix} = \begin{bmatrix} a_{11}x_1 + \cdots + a_{1n}x_n \\ \cdots \\ a_{m1}x_1 + \cdots + a_{mn}x_n \end{bmatrix}.$$

The condition $A\mathbf{v} = \mathbf{0}$ is thus equivalent to the set of simultaneous linear equations

$$\text{(10.7)} \qquad \begin{cases} a_{11}x_1 + \cdots + a_{1n}x_n = 0 \\ \cdots \\ a_{m1}x_1 + \cdots + a_{mn}x_n = 0. \end{cases}$$

This system of m simultaneous equations is surely consistent, since it always has the trivial solution $x_1 = 0, \ldots, x_n = 0$. The rank of its matrix of coefficients is r, by hypothesis. Applying Theorem 9.7 (or the discussion in Section 7), it follows that we can solve for r of the variables $x_1, \ldots, x_n$ in terms of the remaining $n-r$ variables, which may be arbitrary.

For simplicity of notation, we continue the proof in the special case where $n = 5$ and $r = 3$. Suppose (for convenience) that we can solve (10.7) for x_1, x_2, x_3 in terms of x_4, x_5, say

$$x_1 = \alpha_1 x_4 + \alpha_2 x_5, \qquad x_2 = \beta_1 x_4 + \beta_2 x_5, \qquad x_3 = \gamma_1 x_4 + \gamma_2 x_5,$$
x_4, x_5 arbitrary.

Then the kernel consists of all vectors

$$\begin{bmatrix} x_1 \\ x_2 \\ x_3 \\ x_4 \\ x_5 \end{bmatrix} = \begin{bmatrix} \alpha_1 x_4 + \alpha_2 x_5 \\ \beta_1 x_4 + \beta_2 x_5 \\ \gamma_1 x_4 + \gamma_2 x_5 \\ x_4 \\ x_5 \end{bmatrix} = x_4 \begin{bmatrix} \alpha_1 \\ \beta_1 \\ \gamma_1 \\ 1 \\ 0 \end{bmatrix} + x_5 \begin{bmatrix} \alpha_2 \\ \beta_2 \\ \gamma_2 \\ 0 \\ 1 \end{bmatrix},$$

with x_4, x_5 arbitrary. If we put

$$\mathbf{w} = \begin{bmatrix} \alpha_1 \\ \beta_1 \\ \gamma_1 \\ 1 \\ 0 \end{bmatrix}, \qquad \mathbf{w}' = \begin{bmatrix} \alpha_2 \\ \beta_2 \\ \gamma_2 \\ 0 \\ 1 \end{bmatrix},$$

then the kernel consists of *all* linear combinations of $\mathbf{w}$ and $\mathbf{w}'$, and hence is the vector space spanned by $\mathbf{w}$ and $\mathbf{w}'$. But these two vectors are linearly independent (neither is a scalar multiple of the other), since $\mathbf{w}$ ends in $\begin{bmatrix} 1 \\ 0 \end{bmatrix}$ and $\mathbf{w}'$ ends in $\begin{bmatrix} 0 \\ 1 \end{bmatrix}$. Hence the kernel is the vector space with basis $\{\mathbf{w}, \mathbf{w}'\}$, so its dimension equals 2.

An analogous argument works in general, and shows that the kernel is a vector space of dimension $n - r$.]*

(10.8) Corollary

For the linear transformation (10.1),

$$\textit{dimension of kernel} + \textit{dimension of range} = n.$$

PROOF: Use (10.5) and (10.6).

(10.9) Corollary

Let A be an $n \times n$ matrix. Then there exists a nonzero vector $\mathbf{v}$ such that $A\mathbf{v} = \mathbf{0}$ if and only if $\det A = 0$.

PROOF: Such a vector $\mathbf{v}$ exists if and only if the null space of A has dimension at least one. But by (10.6) this dimension is $n - r$, where r is the rank of A. Hence $\mathbf{v}$ exists if and only if $r < n$. By Theorem 9.4, $r < n$ if and only if $\det A = 0$. This completes the proof. (We might equally well have proved this result by using Theorem 7.13.)

EXAMPLES

1. $A = \begin{bmatrix} 1 & -1 \\ 2 & 0 \\ 1 & 2 \end{bmatrix}^{3\times 2}$. The columns of A are linearly independent and are therefore a basis for the column space of A. Thus A has rank 2, and the range of the linear transformation $\mathbf{v} \to A\mathbf{v}$ is a 2-dimensional vector space having the columns of A as basis. Let us find the kernel of the linear transformation. Write

$$A\mathbf{v} = \begin{bmatrix} 1 & -1 \\ 2 & 0 \\ 1 & 2 \end{bmatrix} \begin{bmatrix} x \\ y \end{bmatrix} = \begin{bmatrix} x-y \\ 2x \\ x+2y \end{bmatrix} = \begin{bmatrix} 0 \\ 0 \\ 0 \end{bmatrix},$$

getting the system of simultaneous equations

$$x - y = 0, \qquad 2x = 0, \qquad x + 2y = 0.$$

The only solution is $x = 0$, $y = 0$, so the kernel consists only of the zero vector, and is a 0-dimensional space. Note that (10.8) checks in this case: $0 + 2 = 2$.

2. Consider the matrix A of Example (10.5a). The rows of A are linearly independent, so A has rank 2. The column space of A is therefore 2-dimensional, and any two linearly independent columns of A form a basis for this column space. We have already seen in (10.5a) that the null space of A is also 2-dimensional. Again (10.8) checks, as expected: $2 + 2 = 4$.

*[We have seen in (9.10)–(9.14) that infinite dimensional vector spaces arise naturally in calculus. Let V, W be a pair of vector spaces whose dimensions may be finite or infinite, and let T: $V \to W$ be a mapping which assigns to each vector $\mathbf{v}$ in V a vector $T(\mathbf{v})$ in W. As in (10.2), we call T a *linear transformation* from V to W if T satisfies the identity

$$T(\alpha\mathbf{v} + \beta\mathbf{v}') = \alpha T(\mathbf{v}) + \beta T(\mathbf{v}')$$

for all vectors $\mathbf{v}$, $\mathbf{v}'$ in V, and all scalars α, β.

(10.10) EXAMPLE

Let V be the vector space consisting of all real-valued functions, and let W be the same space as V. Define T: $V \to W$ by setting $T(f) = 3f$ for each function f in V. Then T is a linear transformation, since if f and g are in V, then

$$T(\alpha f + \beta g) = 3(\alpha f + \beta g) = \alpha \cdot 3f + \beta \cdot 3g = \alpha T(f) + \beta T(g).$$

* The remainder of this section is optional material.

(10.11) EXAMPLE

Let V be the vector space consisting of all real-valued functions, and let R consist of all real numbers. Define $T: V \to R$ by mapping each function f in V onto its value $f(1)$. Thus $f \xrightarrow{T} f(1)$, that is, $T(f) = f(1)$. Therefore

$$\begin{aligned}(\alpha f + \beta g) \xrightarrow{T} \text{value of } \alpha f + \beta g \text{ at } 1 &= \alpha f(1) + \beta g(1)\\ &= \alpha T(f) + \beta T(g).\end{aligned}$$

This shows that $T(\alpha f + \beta g) = \alpha T(f) + \beta T(g)$, so that T is a linear transformation.

(10.12) EXAMPLE

Let V be the set of all functions f which are defined and have a derivative at every point of the closed interval $[0,1]$. Let W be the set of all functions defined at every point of $[0,1]$. Define $D: V \to W$ by $f \xrightarrow{D} f'$, that is, $D(f) = f'$. Then D is a linear transformation of the vector space V into the vector space W, since

$$D(\alpha f + \beta g) = (\alpha f + \beta g)' = \alpha f' + \beta g' = \alpha Df + \beta Dg.$$

The kernel of D consists of all functions f in V such that $D(f) = 0$, that is, such that f' is the zero function. Such an f must then be a constant function, and the kernel of D is precisely the set of constant functions. This kernel is then a 1-dimensional vector space contained in V.

(10.13) EXAMPLE

Let $V = C[0,1]$, the vector space consisting of all functions defined and continuous at every point of the closed interval $[0,1]$. Let R be the vector space consisting of all real numbers. Define a mapping $T: V \to R$ by setting

$$T(f) = \int_0^1 f(x)\, dx$$

for every function f in V. Then T is a linear transformation, since

$$\begin{aligned}T(\alpha f + \beta g) = \int_0^1 (\alpha f(x) + \beta g(x))\, dx &= \alpha \int_0^1 f(x)\, dx + \beta \int_0^1 g(x)\, dx\\ &= \alpha T(f) + \beta T(g).\end{aligned}$$

The kernel of T is the vector space S defined in Example 9.14.]*

EXERCISES

1. Find a linear transformation such that

$$\begin{bmatrix}1\\0\end{bmatrix} \longrightarrow \begin{bmatrix}3\\1\end{bmatrix}, \quad \begin{bmatrix}0\\1\end{bmatrix} \longrightarrow \begin{bmatrix}-1\\1\end{bmatrix}.$$

2. Find a linear transformation such that

$$\begin{bmatrix}1\\2\end{bmatrix} \longrightarrow \begin{bmatrix}1\\1\end{bmatrix}, \quad \begin{bmatrix}2\\3\end{bmatrix} \longrightarrow \begin{bmatrix}4\\0\end{bmatrix}.$$

3. Find a linear transformation such that

$$\begin{bmatrix}-1\\1\\1\end{bmatrix} \longrightarrow \begin{bmatrix}2\\1\\4\end{bmatrix}, \quad \begin{bmatrix}0\\1\\2\end{bmatrix} \longrightarrow \begin{bmatrix}3\\2\\0\end{bmatrix}, \quad \begin{bmatrix}0\\4\\3\end{bmatrix} \longrightarrow \begin{bmatrix}1\\1\\1\end{bmatrix}.$$

4. Find the kernel and range of each of the above linear transformations.
5. Find the kernel and range of the linear transformation $\mathbf{v} \rightarrow A\mathbf{v}$, where

$$A = \begin{bmatrix}1 & 2 & 1 & 1 & 0\\1 & -1 & 2 & 0 & 1\end{bmatrix}.$$

6. Let V be a vector space with basis $\{\mathbf{v}_1, \ldots, \mathbf{v}_k\}$, and let F be a linear transformation from V into a vector space W [that is, F satisfies the identity (10.3)]. Show how to compute $F(\mathbf{v})$ for any vector $\mathbf{v}$ in V, once $F(\mathbf{v}_1), \ldots, F(\mathbf{v}_k)$ are known.
7. Let V be the vector space with basis $\mathbf{v}_1 = \begin{bmatrix}1\\0\\0\end{bmatrix}$, $\mathbf{v}_2 = \begin{bmatrix}0\\1\\0\end{bmatrix}$, and let W be the vector space with basis $\mathbf{w}_2 = \begin{bmatrix}0\\1\\0\end{bmatrix}$, $\mathbf{w}_3 = \begin{bmatrix}0\\0\\1\end{bmatrix}$. Find a linear transformation F from V to W such that

$$F(\mathbf{v}_1) = \mathbf{w}_2 + \mathbf{w}_3, \qquad F(\mathbf{v}_2) = \mathbf{w}_2 - \mathbf{w}_3.$$

What is $F\left(\begin{bmatrix}\alpha\\\beta\\0\end{bmatrix}\right)$?

*8. If $B^{m \times m}$ is nonsingular and A is any $m \times n$ matrix, show that BA has the same null space as A. Then use Theorem 10.6 to prove that BA has the same rank as A.

*9. Deduce from the previous exercise that if $C^{n\times n}$ is nonsingular, then A and AC have the same rank. [*Hint:* Use transposes.]

*10. Let V and W both be the vector space consisting of all real polynomials of at most degree 3. Define D: $V \to W$ by $D(f) = f'$, for each f in V. Show that D is a linear transformation. Determine the kernel of D and the range of D, and verify that (10.8) holds true in this case.

*11. Let V, W be vector spaces, and let F and G be a pair of linear transformations from V to W. Prove that $F + G$ is also a linear transformation, where (by definition)

$$(F + G)(\mathbf{v}) = F(\mathbf{v}) + G(\mathbf{v}).$$

*12. Let V be the set of all real-valued functions f defined and differentiable at every point of the closed interval $[0,1]$. Let R be the vector space consisting of all real numbers. Define T: $V \to R$ by

$$T(f) = f(0) + f'(0) + 3f'(1).$$

Prove that T is a linear transformation.

*13. Let V, R be as in Exercise 12. Define a mapping ϕ: $V \to R$ by setting $\phi(f) = \{f(0)\}^2$. Show that ϕ is *not* a linear transformation.

11 CHARACTERISTIC ROOTS AND CHARACTERISTIC VECTORS

Let $A = [a_{ij}]^{n\times n}$ be a square matrix, and let λ denote a variable. Form the $n \times n$ matrix

$$\text{(11.1)} \qquad \lambda I - A = \begin{bmatrix} \lambda - a_{11} & -a_{12} & \cdots & -a_{1n} \\ -a_{21} & \lambda - a_{22} & \cdots & -a_{2n} \\ \cdots & \cdots & \cdots & \cdots \\ -a_{n1} & -a_{n2} & \cdots & \lambda - a_{nn} \end{bmatrix}.$$

We set

$$f(\lambda) = \det (\lambda I - A) = \textit{characteristic polynomial } \text{of } A.$$

Then $f(\lambda)$ is a polynomial of degree n in the variable λ, and the leading term is λ^n. The constant term of the polynomial $f(\lambda)$ is obtained by setting $\lambda = 0$ in the expression for $f(\lambda)$. Thus the constant term is $\det(-A)$, and this equals $(-1)^n \det A$.

EXAMPLES

1. If $A = [a_{ij}]^{2\times 2}$, then

$$f(\lambda) = \begin{vmatrix} \lambda - a_{11} & -a_{12} \\ -a_{21} & \lambda - a_{22} \end{vmatrix} = (\lambda - a_{11})(\lambda - a_{22}) - a_{21}a_{12}$$
$$= \lambda^2 - (a_{11} + a_{22})\lambda + (a_{11}a_{22} - a_{21}a_{12}).$$

2. If $A = [a_{ij}]^{3\times 3}$, then

$$f(\lambda) = \begin{vmatrix} \lambda - a_{11} & -a_{12} & -a_{13} \\ -a_{21} & \lambda - a_{22} & -a_{23} \\ -a_{31} & -a_{32} & \lambda - a_{33} \end{vmatrix}$$
$$= \lambda^3 - (a_{11} + a_{22} + a_{33})\lambda^2$$
$$+ (a_{11}a_{22} - a_{12}a_{21} + a_{11}a_{33} - a_{13}a_{31} + a_{22}a_{33} - a_{23}a_{32})\lambda - \det A.$$

In general, we obtain

$$\text{(11.2)} \quad f(\lambda) = \lambda^n - (a_{11} + a_{22} + \cdots + a_{nn})\lambda^{n-1} + \cdots + (-1)^n \det A.$$

The expression $a_{11} + a_{22} + \cdots + a_{nn}$ is the sum of the main diagonal entries of A, and is called the *trace* of the matrix A. This trace is then a number associated with a square matrix; the concept is useful in certain parts of matrix theory.

The *characteristic equation* of A is $f(\lambda) = 0$, and its roots are the *characteristic roots* of the matrix A (also called the *eigenvalues* of A). An $n \times n$ matrix A has n characteristic roots, $\lambda_1, \ldots, \lambda_n$, not necessarily distinct from one another. We may write

(11.3) $$\det(\lambda I - A) = f(\lambda) = (\lambda - \lambda_1)(\lambda - \lambda_2) \cdots (\lambda - \lambda_n).$$

The characteristic roots of A are the values of λ for which $f(\lambda) = 0$, that is, for which $\det(\lambda I - A) = 0$. *Therefore the matrix $\lambda I - A$ is singular if and only if λ is a characteristic root of A.*

EXAMPLES

1. $$A = \begin{bmatrix} 2 & 1 \\ 3 & 2 \end{bmatrix}, \quad f(\lambda) = \begin{vmatrix} \lambda - 2 & -1 \\ -3 & \lambda - 2 \end{vmatrix} = \lambda^2 - 4\lambda + 1.$$

The characteristic roots of A are the solutions of the characteristic equation $\lambda^2 - 4\lambda + 1 = 0$, that is,

$$\lambda = \frac{4 \pm \sqrt{16 - 4}}{2} = 2 \pm \sqrt{3}.$$

(For a 2×2 matrix A, of course

$$f(\lambda) = \lambda^2 - (\text{trace of } A)\lambda + \det A,$$

so we could have written down the characteristic equation directly.)

2. $$A = \begin{bmatrix} 1 & 2 & 3 \\ 0 & 2 & 4 \\ 0 & 0 & 6 \end{bmatrix},$$

$$f(\lambda) = \begin{vmatrix} \lambda - 1 & -2 & -3 \\ 0 & \lambda - 2 & -4 \\ 0 & 0 & \lambda - 6 \end{vmatrix} = (\lambda - 1)(\lambda - 2)(\lambda - 6),$$

and A has characteristic roots 1, 2, and 6.

Now let λ be a characteristic root of $A^{n \times n}$. Then $\det(\lambda I - A) = 0$, so by (10.9) there exists a nonzero vector $\mathbf{v}$ such that

$$(\lambda I - A)\mathbf{v} = \mathbf{0}.$$

We may rewrite this equation as $(\lambda I)\mathbf{v} - A\mathbf{v} = \mathbf{0}$; but $(\lambda I)\mathbf{v} = \lambda\mathbf{v}$, so we obtain

(11.4) $$A\mathbf{v} = \lambda\mathbf{v}.$$

Conversely, suppose that $\mathbf{v}$ is a nonzero vector such that (11.4)

holds true for some scalar λ. Retracing our steps, we see then that $(\lambda I - A)\mathbf{v} = \mathbf{0}$. But by (10.9), if $(\lambda I - A)\mathbf{v} = \mathbf{0}$ for some nonzero vector $\mathbf{v}$, then $\det(\lambda I - A) = 0$, and so λ must be a characteristic root of A.

We have therefore proved

(11.5) Theorem

If λ is a characteristic root of A, then there is at least one nonzero vector $\mathbf{v}$ such that $A\mathbf{v} = \lambda\mathbf{v}$. Conversely, if $\mathbf{v}$ is a nonzero vector satisfying $A\mathbf{v} = \lambda\mathbf{v}$ for some scalar λ, then this value of λ must be a characteristic root of A.

Let λ be a characteristic root of A, and let $\mathbf{v}$ be a *nonzero* vector such that $A\mathbf{v} = \lambda\mathbf{v}$. We shall say that $\mathbf{v}$ is a *characteristic vector* (or *eigenvector*) of A belonging to the characteristic root λ.

EXAMPLES

1. (See Example 2 above.) Let $A = \begin{bmatrix} 1 & 2 & 3 \\ 0 & 2 & 4 \\ 0 & 0 & 6 \end{bmatrix}$. We have

$$\begin{bmatrix} 1 & 2 & 3 \\ 0 & 2 & 4 \\ 0 & 0 & 6 \end{bmatrix} \begin{bmatrix} 2 \\ 1 \\ 0 \end{bmatrix} = \begin{bmatrix} 4 \\ 2 \\ 0 \end{bmatrix} = 2 \begin{bmatrix} 2 \\ 1 \\ 0 \end{bmatrix}.$$

Thus $\begin{bmatrix} 2 \\ 1 \\ 0 \end{bmatrix}$ is a characteristic vector of A belonging to the characteristic root 2.

Let us find a characteristic vector $\mathbf{v}$ belonging to the characteristic root 6 of A. We must solve $A\mathbf{v} = 6\mathbf{v}$, that is, $(6I - A)\mathbf{v} = \mathbf{0}$. This equation gives

$$(6I - A)\mathbf{v} = \begin{bmatrix} 5 & -2 & -3 \\ 0 & 4 & -4 \\ 0 & 0 & 0 \end{bmatrix} \begin{bmatrix} x \\ y \\ z \end{bmatrix} = \begin{bmatrix} 5x - 2y - 3z \\ 4y - 4z \\ 0 \end{bmatrix} = \begin{bmatrix} 0 \\ 0 \\ 0 \end{bmatrix},$$

that is,

$$5x - 2y - 3z = 0, \qquad 4y - 4z = 0, \qquad 0 = 0.$$

Solving these simultaneously, we obtain

$$x = z, \qquad y = z, \quad z \text{ arbitrary}.$$

Thus

$$\mathbf{v} = \begin{bmatrix} z \\ z \\ z \end{bmatrix}, \quad z \text{ arbitrary}.$$

Taking $z = 1$, we obtain a characteristic vector $\begin{bmatrix} 1 \\ 1 \\ 1 \end{bmatrix}$ of A belonging to the characteristic root 6. Note that in this case, the only possible characteristic vectors belonging to 6 are nonzero scalar multiples of $\begin{bmatrix} 1 \\ 1 \\ 1 \end{bmatrix}$.

2. Let $A = \begin{bmatrix} 0 & 4 \\ 0 & 0 \end{bmatrix}$. The characteristic roots of A are 0, 0. Let us find all characteristic vectors $\mathbf{v}$ belonging to the root 0. Condition (11.4) becomes $A\mathbf{v} = \mathbf{0}$, that is,

$$\begin{bmatrix} 0 & 4 \\ 0 & 0 \end{bmatrix} \begin{bmatrix} x \\ y \end{bmatrix} = \begin{bmatrix} 4y \\ 0 \end{bmatrix} = \begin{bmatrix} 0 \\ 0 \end{bmatrix}.$$

This gives $y = 0$, x arbitrary, so $\mathbf{v} = \begin{bmatrix} x \\ 0 \end{bmatrix}$. Taking $x = 1$, we get the characteristic vector $\begin{bmatrix} 1 \\ 0 \end{bmatrix}$ belonging to the characteristic root 0. Every other characteristic vector is a nonzero scalar multiple of $\begin{bmatrix} 1 \\ 0 \end{bmatrix}$, even though 0 occurs as a characteristic root with multiplicity 2.

3. Let $A = \begin{bmatrix} 1 & 0 \\ 0 & 1 \end{bmatrix}$; the characteristic roots are 1, 1. The characteristic vectors belonging to the root 1 are easily found to be $\mathbf{v} = \begin{bmatrix} x \\ y \end{bmatrix}$, x, y arbitrary (but not both zero). In this case, there are two linearly independent characteristic vectors $\begin{bmatrix} 1 \\ 0 \end{bmatrix}$ and $\begin{bmatrix} 0 \\ 1 \end{bmatrix}$, both belonging to the characteristic root 1 of A.

Let us study a bit more carefully the set of vectors $\mathbf{v}$ satisfying (11.4). Of course $\mathbf{v} = \mathbf{0}$ is a solution of (11.4), although $\mathbf{0}$ is not counted as a characteristic vector of A. The nonzero vectors $\mathbf{v}$ for which (11.4) holds are (by definition) the characteristic vectors of A belonging to λ. We now prove

(11.6) Theorem

Let λ be a characteristic root of A, and let V be the set of all vectors $\mathbf{v}$ such that $A\mathbf{v} = \lambda\mathbf{v}$. Then V is a vector space, so that any linear combination of vectors in V again lies in V.

PROOF: We have seen that $A\mathbf{v} = \lambda\mathbf{v}$ if and only if $(\lambda I - A)\mathbf{v} = \mathbf{0}$. Thus V consists of all vectors $\mathbf{v}$ such that $(\lambda I - A)\mathbf{v} = \mathbf{0}$. This means that V is the null space[1] of the matrix $\lambda I - A$. Hence by Theorem 10.6, V is a vector space. It then follows from (9.1a) that every linear combination of vectors in V again lies in V. This completes the proof.

We may list two consequences of this theorem:

(i) If $\mathbf{v}$ is a characteristic vector of A belonging to the characteristic root λ, and if c is a nonzero scalar, then $c\mathbf{v}$ also belongs to λ.
(ii) If $\mathbf{v}$ and $\mathbf{v}'$ are characteristic vectors of A belonging to the *same* characteristic root λ, and if c, c' are scalars such that $c\mathbf{v} + c'\mathbf{v}' \neq \mathbf{0}$, then $c\mathbf{v} + c'\mathbf{v}'$ is also a characteristic vector of A belonging to λ.

EXAMPLE

Let $A = \begin{bmatrix} 2 & 0 & 3 \\ 0 & 2 & 0 \\ 0 & 0 & 5 \end{bmatrix}$; its characteristic roots are 2, 2, and 5. Let us find all characteristic vectors of A belonging to 2. We first determine the null space of $2I - A$, by solving $(2I - A)\mathbf{v} = \mathbf{0}$. This gives

$$(2I - A)\mathbf{v} = \begin{bmatrix} 0 & 0 & -3 \\ 0 & 0 & 0 \\ 0 & 0 & -3 \end{bmatrix}\begin{bmatrix} x \\ y \\ z \end{bmatrix} = \begin{bmatrix} -3z \\ z \\ -3z \end{bmatrix} = \begin{bmatrix} 0 \\ 0 \\ 0 \end{bmatrix}.$$

Hence $z = 0$, while x and y are arbitrary, so

$$\mathbf{v} = \begin{bmatrix} x \\ y \\ 0 \end{bmatrix} = x\begin{bmatrix} 1 \\ 0 \\ 0 \end{bmatrix} + y\begin{bmatrix} 0 \\ 1 \\ 0 \end{bmatrix}.$$

Thus $\mathbf{v}_1 = \begin{bmatrix} 1 \\ 0 \\ 0 \end{bmatrix}$ and $\mathbf{v}_2 = \begin{bmatrix} 0 \\ 1 \\ 0 \end{bmatrix}$ are a pair of linearly independent characteristic vectors belonging to the characteristic root 2. These vectors are a basis for the null space of $2I - A$. The above vector $\mathbf{v}$ satisfies $A\mathbf{v} = 2\mathbf{v}$, and $\mathbf{v}$ is a characteristic vector if $\mathbf{v} \neq \mathbf{0}$, that is, if x and y are not both zero.

Let us give a geometric interpretation of characteristic vectors. Consider the linear transformation of the XY-plane into itself, which maps the point (x,y) onto the point $(3x + y,\ x + 3y)$. The matrix associated with this transformation is

$$A = \begin{bmatrix} 3 & 1 \\ 1 & 3 \end{bmatrix},$$

and if $\mathbf{v} = \begin{bmatrix} x \\ y \end{bmatrix}$, then A maps $\mathbf{v}$ onto $A\mathbf{v}$. There are two characteristic roots

[1] Recall that the *null space* of a matrix B is the set of vectors $\mathbf{v}$ such that $B\mathbf{v} = \mathbf{0}$.

of A, namely 2, 4. Let $\mathbf{v}_1 = \begin{bmatrix} 1 \\ -1 \end{bmatrix}$, $\mathbf{v}_2 = \begin{bmatrix} 1 \\ 1 \end{bmatrix}$. It is easily checked that $A\mathbf{v}_1 = 2\mathbf{v}_1$, $A\mathbf{v}_2 = 4\mathbf{v}_2$, so $\mathbf{v}_1$ belongs to 2, and $\mathbf{v}_2$ belongs to 4. (See Figure 11.1.)

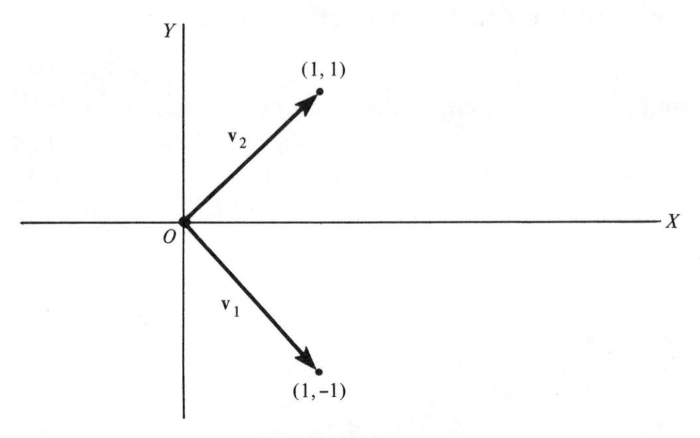

Figure 11.1

We can visualize the effect of the transformation A thus: it stretches all displacements in the $\mathbf{v}_1$ direction by a factor of 2, and all displacements in the direction $\mathbf{v}_2$ by a factor of 4. (It is a coincidence that in this case $\mathbf{v}_1$ and $\mathbf{v}_2$ are perpendicular to each other.)

*[For certain types of matrices, we can make some general assertions about their characteristic roots. In proving these results, we need some facts about complex numbers. Let $c = a + bi$, a, b real; define its *complex conjugate* $\bar{c}$ by the equation $\bar{c} = a - bi$. Then

$$|c|^2 = \bar{c}c = a^2 + b^2,$$

and the *absolute value* $|c|$ is not zero unless $c = 0$. For a matrix A, we may form $\bar{A}$ by replacing each entry of A by its complex conjugate. If all entries of A are real numbers, we call A a *real* matrix. Then A is real if and only if $A = \bar{A}$.

From the formulas

$$\overline{c + c'} = \bar{c} + \bar{c}', \qquad \overline{cc'} = \bar{c} \cdot \bar{c}',$$

it follows that for any matrices A, B, C

$$\overline{A + B} = \bar{A} + \bar{B}, \qquad \overline{AC} = \bar{A}\,\bar{C}.$$

(11.7) Theorem

Each characteristic root of a real symmetrix matrix is real.

PROOF: Let A be a real symmetric matrix. Then $A = \bar{A}$ since A is real, and $A = A^{\mathrm{T}}$ since A is symmetric. Combining these two facts, we

* The remainder of this section is optional material.

obtain $A = \bar{A}^{\mathrm{T}}$. Now let λ be a characteristic root of A. We shall prove that λ is real by showing that $\lambda = \bar{\lambda}$.

By (11.5), there exists a characteristic vector $\mathbf{v}$ belonging to λ, that is, $A\mathbf{v} = \lambda\mathbf{v}$. Therefore $(A\mathbf{v})^{\mathrm{T}} = (\lambda\mathbf{v})^{\mathrm{T}}$, which we may write as $\mathbf{v}^{\mathrm{T}}A^{\mathrm{T}} = \lambda\mathbf{v}^{\mathrm{T}}$. Taking complex conjugates, we obtain

$$\bar{\mathbf{v}}^{\mathrm{T}}\bar{A}^{\mathrm{T}} = \bar{\lambda}\bar{\mathbf{v}}^{\mathrm{T}}.$$

Since $A = \bar{A}^{\mathrm{T}}$, this last equation may be written as

$$\bar{\mathbf{v}}^{\mathrm{T}}A = \bar{\lambda}\bar{\mathbf{v}}^{\mathrm{T}}.$$

Multiplying on the right by $\mathbf{v}$, we get

$$(\bar{\mathbf{v}}^{\mathrm{T}}A)\mathbf{v} = \bar{\lambda}\bar{\mathbf{v}}^{\mathrm{T}}\mathbf{v}.$$

On the other hand, since $A\mathbf{v} = \lambda\mathbf{v}$, we have also

$$\bar{\mathbf{v}}^{\mathrm{T}}(A\mathbf{v}) = \bar{\mathbf{v}}^{\mathrm{T}}(\lambda\mathbf{v}) = \lambda\bar{\mathbf{v}}^{\mathrm{T}}\mathbf{v}.$$

Since

$$(\bar{\mathbf{v}}^{\mathrm{T}}A)\mathbf{v} = \bar{\mathbf{v}}^{\mathrm{T}}(A\mathbf{v}),$$

we obtain

(11.8) $$\bar{\lambda}(\bar{\mathbf{v}}^{\mathrm{T}}\mathbf{v}) = \lambda(\bar{\mathbf{v}}^{\mathrm{T}}\mathbf{v}).$$

But if $c_1, \ldots, c_n$ are the entries of $\mathbf{v}$, then

$$\bar{\mathbf{v}}^{\mathrm{T}}\mathbf{v} = [\bar{c}_1 \quad \cdots \quad \bar{c}_n]\begin{bmatrix} c_1 \\ \cdot \\ \cdot \\ \cdot \\ c_n \end{bmatrix} = |c_1|^2 + \cdots + |c_n|^2.$$

Since $\mathbf{v} \neq \mathbf{0}$, surely $|c_1|^2 + \cdots + |c_n|^2 \neq 0$, so (11.8) gives $\lambda = \bar{\lambda}$. Therefore λ is real, and the proof is complete.

Remark: As a matter of fact, the same proof would hold even if A were not a real matrix, provided A satisfied the equation

$$A = \bar{A}^{\mathrm{T}}.$$

Such matrices are called *Hermitian* matrices. We have just shown that the characteristic roots of Hermitian matrices are all real.

EXAMPLE

The matrix $A = \begin{bmatrix} 2 & 1+i \\ 1-i & 1 \end{bmatrix}$ is Hermitian. Its characteristic equation is $\lambda^2 - 3\lambda = 0$, with roots 0, 3.

In a similar vein, we prove

(11.9) Theorem

Let A be a real skew-symmetric matrix. Each of its characteristic roots is either zero or pure imaginary.

PROOF: The hypothesis implies that $A = -\bar{A}^{\mathrm{T}}$. If $A\mathbf{v} = \lambda\mathbf{v}$, then we find that

$$\bar{\mathbf{v}}^{\mathrm{T}}A = -\bar{\lambda}\bar{\mathbf{v}}^{\mathrm{T}}.$$

Evaluating $\bar{\mathbf{v}}^{\mathrm{T}}A\mathbf{v}$ in two ways, as in (11.6), we get

$$-\bar{\lambda}(\bar{\mathbf{v}}^{\mathrm{T}}\mathbf{v}) = \lambda(\bar{\mathbf{v}}^{\mathrm{T}}\mathbf{v}),$$

so $\lambda = -\bar{\lambda}$. Therefore the real part of λ is zero, so either $\lambda = 0$ or λ is pure imaginary, as claimed.

The following result is often useful (see references for proof).

(11.10) Theorem

Let $\lambda_1, \ldots, \lambda_n$ be the characteristic roots of A. Then for each positive integer k, the characteristic roots of A^k are $\lambda_1{}^k, \ldots, \lambda_n{}^k$ This also holds when k is a negative integer, provided that A is nonsingular. In general, the characteristic roots of the matrix

$$\alpha_0 A^k + \alpha_1 A^{k-1} + \cdots + \alpha_{k-1}A + \alpha_k I$$

are

$$\{\alpha_0\lambda_i{}^k + \alpha_1\lambda_i{}^{k-1} + \cdots + \alpha_{k-1}\lambda_i + \alpha_k, \quad i = 1, \ldots, n\}.]^*$$

EXERCISES

1. Find characteristic roots and characteristic vectors for each of the following matrices:

$$\begin{bmatrix} 1 & 2 & 3 \\ 0 & 2 & 4 \\ 0 & 0 & 2 \end{bmatrix}, \begin{bmatrix} 1 & 1 \\ -1 & 3 \end{bmatrix}, \begin{bmatrix} 0 & 1 \\ 1 & 0 \end{bmatrix}, \begin{bmatrix} 0 & 1 \\ -1 & 0 \end{bmatrix}, \begin{bmatrix} a & 0 \\ 0 & b \end{bmatrix}, \begin{bmatrix} 2 & 0 \\ 1 & 2 \end{bmatrix},$$

$$\begin{bmatrix} -1 & 0 & 0 & 0 \\ 0 & 0 & 0 & 0 \\ 0 & 0 & 0 & 1 \\ 0 & 0 & 1 & 0 \end{bmatrix}, \begin{bmatrix} 0 & 0 & 0 & 1 \\ 0 & 0 & 1 & 0 \\ 0 & 1 & 0 & 0 \\ 1 & 0 & 0 & 0 \end{bmatrix}, \begin{bmatrix} 0 & 1 \\ -1 & -1 \end{bmatrix}, \begin{bmatrix} \cos\theta & \sin\theta \\ -\sin\theta & \cos\theta \end{bmatrix}.$$

2. Find x, y, z not all zero, such that for some constant c,

$$\begin{cases} 2x + y - z = cx \\ \quad\quad\; y - z = cy \\ \quad\;\; 8y - 5z = cz. \end{cases}$$

3. If $A = [a_{ij}]^{3\times 3}$, show that the coefficient of λ in the characteristic equation of A is

$$\begin{vmatrix} a_{11} & a_{12} \\ a_{21} & a_{22} \end{vmatrix} + \begin{vmatrix} a_{11} & a_{13} \\ a_{31} & a_{33} \end{vmatrix} + \begin{vmatrix} a_{22} & a_{23} \\ a_{32} & a_{33} \end{vmatrix}.$$

*4. Comparing equations (11.2) and (11.3), prove that

trace of A = sum of the characteristic roots of A,
det A = product of the characteristic roots of A.

*5. Let λ be a characteristic root of $A^{n\times n}$, and let r be the rank of the matrix $\lambda I - A$. Show that there are $n - r$ linearly independent characteristic vectors of A which belong to the root λ, but that there cannot be more than $n - r$ such vectors.

*6. Given $A^{m\times m}$, $B^{n\times n}$, let $C = \begin{bmatrix} A^{m\times m} & 0^{m\times n} \\ 0^{n\times m} & B^{n\times n} \end{bmatrix}$. Show that the characteristic polynomial of C is the product of the characteristic polynomials of A and B. What is the relation between the characteristic roots of A, B, and C? What is the relation between their characteristic vectors?

7. Let

$$A = \begin{bmatrix} a & b \\ c & d \end{bmatrix}, \qquad f(\lambda) = \lambda^2 - (a + d)\lambda + (ad - bc) = \det(\lambda I - A).$$

Show that the matrix $f(A)$ is the zero matrix, where

$$f(A) = A^2 - (a + d)A + (ad - bc)I.$$

(This is a special case of the *Hamilton-Cayley Theorem: Every matrix satisfies its own characteristic equation.*)

8. What are the characteristic roots of a diagonal matrix? Verify the validity of Theorem 11.9 in this case.

9. Show that $A^{n\times n}$ is singular if and only if 0 is a characteristic root of A.

10. Let

$$x' = 2x + 3y, \qquad y' = 3y$$

define a mapping of $P(x,y)$ onto the point $P'(x',y')$ in two-dimensional space. As P ranges over all points on the x-axis, what points P' are obtained? As P ranges over all points on the y-axis, what about P'? As P ranges over the points on the line $x + y = 7$, what about P'? For which points P is it true that $\overrightarrow{OP'} = \lambda \cdot \overrightarrow{OP}$ for some λ? Use your answer to give a geometric description of the mapping $P \to P'$.

11. Answer the questions in the preceding problem for the mapping given by

$$x' = -2x + y, \qquad y' = -3x.$$

*12. Call a complex matrix A *skew-Hermitian* if $\bar{A}^T = -A$. Show that the characteristic roots of such a matrix are pure imaginary or zero.

13. Show that $(\lambda I - A)^T = \lambda I - A^T$. Deduce from this that A and A^T have the same characteristic polynomial.

14. For what values of λ does there exist a nonzero $1 \times m$ vector $\mathbf{w}$ such that $\mathbf{w}A = \lambda\mathbf{w}$, where A is a given $m \times m$ matrix?

15. Discuss the solution of the system of simultaneous equations

$$\begin{cases} (a - \lambda)x + by = 0 \\ cx + (d - \lambda)y = 0 \end{cases}$$

in the unknowns x, y.

*16. Let A, B be $n \times n$ matrices which commute with each other (that is, $AB = BA$), and let $\mathbf{v}$ be a characteristic vector of A belonging to λ. If $B\mathbf{v} \neq \mathbf{0}$, show that $B\mathbf{v}$ is also a characteristic vector of A belonging to λ.

*17. Let $P^{n\times n}$ be nonsingular, and let $A^{n\times n}$ be arbitrary. Show that

$$\lambda I - P^{-1}AP = P^{-1}(\lambda I - A)P,$$

and that

$$\det(\lambda I - P^{-1}AP) = \det(\lambda I - A).$$

Deduce from this that $P^{-1}AP$ has the same characteristic polynomial as A.

18. Find a matrix $A^{2\times 2}$ with characteristic polynomial $\lambda^2 + \lambda - 1$.

19. Let A, B be 2×2 matrices. Show that AB and BA have the same characteristic polynomial. [*Hint:* It suffices to show that they have the same trace, and the same determinant.]

12 ORTHOGONAL VECTORS

Let us review some ideas concerning vectors in a 3-dimensional Euclidean space. A *unit vector* is a vector of length 1. Let $\mathbf{i}$, $\mathbf{j}$, $\mathbf{k}$ be unit vectors along the coordinate axes in XYZ-space. An arbitrary vector $\mathbf{a}$ in this space may be expressed as

$$\mathbf{a} = a_1\mathbf{i} + a_2\mathbf{j} + a_3\mathbf{k}, \quad a_1,\ a_2,\ a_3 \text{ real numbers.}$$

(In fact, if $\mathbf{a} = \overrightarrow{OP}$ where O is the origin, then the coefficients a_1, a_2, a_3 are just the coordinates of the point P.) The *length* (or *magnitude*) of the vector $\mathbf{a}$ is denoted by $|\mathbf{a}|$, and is given by the formula

$$|\mathbf{a}| = (a_1^2 + a_2^2 + a_3^2)^{1/2}.$$

This length is positive except when $\mathbf{a}$ is the zero vector.

For any positive real number λ, the vector $\lambda\mathbf{a}$ goes in the same direction as $\mathbf{a}$, and its length is λ times the length of $\mathbf{a}$. In particular when $\mathbf{a} \neq \mathbf{0}$, if we choose $\lambda = \dfrac{1}{|\mathbf{a}|}$ we obtain a unit vector $\dfrac{1}{|\mathbf{a}|}\mathbf{a}$ in the direction of $\mathbf{a}$. The procedure of replacing a vector by a unit vector in the same direction is called *normalization,* and we say that we *normalize* $\mathbf{a}$ when we find the unit vector $\dfrac{1}{|\mathbf{a}|}\mathbf{a}$.

Given another vector

$$\mathbf{b} = b_1\mathbf{i} + b_2\mathbf{j} + b_3\mathbf{k},$$

let θ denote the angle between $\mathbf{a}$ and $\mathbf{b}$, measured so that $0° \leqslant \theta \leqslant 180°$. Define the *dot product* (or *inner product*) by

(12.1)
$$\mathbf{a} \cdot \mathbf{b} = |\mathbf{a}||\mathbf{b}| \cos \theta,$$

a scalar quantity. (If either $\mathbf{a}$ or $\mathbf{b}$ is the zero vector, the angle θ is not defined, and we just set $\mathbf{a} \cdot \mathbf{b} = 0$.) We note that

$$\mathbf{i} \cdot \mathbf{i} = \mathbf{j} \cdot \mathbf{j} = \mathbf{k} \cdot \mathbf{k} = 1, \qquad \mathbf{i} \cdot \mathbf{j} = \mathbf{j} \cdot \mathbf{k} = \mathbf{k} \cdot \mathbf{i} = 0,$$

and also $\mathbf{i} \cdot (-\mathbf{i}) = -1$, and so on. Furthermore,

(12.2) $$\mathbf{a} \cdot \mathbf{a} = |\mathbf{a}||\mathbf{a}| \cos 0° = |\mathbf{a}|^2.$$

The following formulas are easily proved by geometric arguments[1]:

(12.3) $$\begin{cases} \mathbf{a} \cdot \mathbf{b} = \mathbf{b} \cdot \mathbf{a}, & \mathbf{a} \cdot (\mathbf{b} + \mathbf{c}) = \mathbf{a} \cdot \mathbf{b} + \mathbf{a} \cdot \mathbf{c}, \\ (\mathbf{a} + \mathbf{b}) \cdot \mathbf{c} = \mathbf{a} \cdot \mathbf{c} + \mathbf{b} \cdot \mathbf{c}, & \mathbf{a} \cdot (\lambda \mathbf{b}) = (\lambda \mathbf{a}) \cdot \mathbf{b} = \lambda(\mathbf{a} \cdot \mathbf{b}). \end{cases}$$

Consequently we obtain

$$\begin{aligned} \mathbf{a} \cdot \mathbf{b} &= (a_1\mathbf{i} + a_2\mathbf{j} + a_3\mathbf{k}) \cdot (b_1\mathbf{i} + b_2\mathbf{j} + b_3\mathbf{k}) \\ &= a_1b_1\, \mathbf{i} \cdot \mathbf{i} + a_1b_2\, \mathbf{i} \cdot \mathbf{j} + \cdots + a_3b_3\, \mathbf{k} \cdot \mathbf{k} \\ &= a_1b_1 + a_2b_2 + a_3b_3. \end{aligned}$$

EXAMPLE

Given two vectors $\mathbf{a} = \mathbf{i} - 2\mathbf{j} + 2\mathbf{k}$, $\mathbf{b} = 4\mathbf{i} - 3\mathbf{k}$, find their lengths and the angle between them. Also find a unit vector in the direction of **a**.

SOLUTION:

$$|\mathbf{a}| = (1 + 4 + 4)^{1/2} = 3, \qquad |\mathbf{b}| = (16 + 9)^{1/2} = 5,$$

unit vector in the direction of $\mathbf{a} = \frac{1}{3}(\mathbf{i} - 2\mathbf{j} + 2\mathbf{k}) = \frac{1}{3}\mathbf{i} - \frac{2}{3}\mathbf{j} + \frac{2}{3}\mathbf{k}$,

$$\mathbf{a} \cdot \mathbf{b} = 1 \cdot 4 - 2 \cdot 0 + 2(-3) = -2,$$

$$\cos\theta = \frac{\mathbf{a} \cdot \mathbf{b}}{|\mathbf{a}||\mathbf{b}|} = \frac{-2}{3 \cdot 5} = \frac{-2}{15}, \qquad \theta = \arccos\left(\frac{-2}{15}\right).$$

We now prove the *Cauchy-Schwartz Inequality:*

(12.4) $$|\mathbf{a} \cdot \mathbf{b}| \leqslant |\mathbf{a}||\mathbf{b}|,$$

that is, the absolute value of the scalar $\mathbf{a} \cdot \mathbf{b}$ is less than or equal to the length of **a** times the length of **b**.

PROOF: If either **a** or **b** equals **0**, then $\mathbf{a} \cdot \mathbf{b} = 0$ and $|\mathbf{a}||\mathbf{b}| = 0$, so equality holds in (12.4). If both **a** and **b** are nonzero, let θ be the angle between them. Since $|\cos\theta| \leqslant 1$, from (12.1) we obtain

$$|\mathbf{a} \cdot \mathbf{b}| = ||\mathbf{a}||\mathbf{b}| \cos\theta| = |\mathbf{a}||\mathbf{b}||\cos\theta| \leqslant |\mathbf{a}||\mathbf{b}|,$$

and the result is proved.

Now we generalize all this to Euclidean n-space: a *vector* is an n-tuple[2] of real numbers $\mathbf{a} = [a_1, \ldots, a_n]$. The *zero vector* is

[1] See references in the Preface.

[2] The commas between $a_1, a_2, \ldots, a_n$ are inserted just for convenience.

$\mathbf{0} = [0, \ldots, 0]$. For any scalar λ, define

$$\lambda\mathbf{a} = \lambda[a_1, \ldots, a_n] = [\lambda a_1, \ldots, \lambda a_n].$$

The *length* (or *magnitude*) of $\mathbf{a}$ is denoted by $|\mathbf{a}|$, and is defined by

(12.5) $$|\mathbf{a}| = (a_1^2 + a_2^2 + \cdots + a_n^2)^{1/2}.$$

This length is positive except when $\mathbf{a} = \mathbf{0}$. Clearly

$$|\lambda\mathbf{a}| = (\lambda^2 a_1^2 + \lambda^2 a_2^2 + \cdots + \lambda^2 a_n^2)^{1/2} = |\lambda||\mathbf{a}|.$$

If $\lambda > 0$, we say that $\lambda\mathbf{a}$ has the "same direction" as $\mathbf{a}$. A *unit vector* is a vector of length 1. If $\mathbf{a} \neq \mathbf{0}$, a unit vector in the same direction as $\mathbf{a}$ is given by

(12.6) $$\frac{1}{|\mathbf{a}|}\mathbf{a} = \left[\frac{a_1}{|\mathbf{a}|}, \frac{a_2}{|\mathbf{a}|}, \ldots, \frac{a_n}{|\mathbf{a}|}\right].$$

Define the *sum* of two vectors by

$$\mathbf{a} + \mathbf{b} = [a_1, \ldots, a_n] + [b_1, \ldots, b_n] = [a_1 + b_1, \ldots, a_n + b_n].$$

There are obvious identities

$$\mathbf{a} + \mathbf{b} = \mathbf{b} + \mathbf{a}, \qquad \lambda(\mathbf{a} + \mathbf{b}) = \lambda\mathbf{a} + \lambda\mathbf{b},$$

and so on. We leave their proofs as exercise for the reader.

Next, we define the *dot product* $\mathbf{a} \cdot \mathbf{b}$ of the vectors $\mathbf{a}$ and $\mathbf{b}$ by the formula

(12.7) $$\mathbf{a} \cdot \mathbf{b} = a_1b_1 + a_2b_2 + \cdots + a_nb_n.$$

Then $\mathbf{a} \cdot \mathbf{b}$ is a scalar quantity, and we leave it to the reader to verify that the identities (12.3) also hold for these vectors in Euclidean n-space. Furthermore we note that

(12.8) $$\mathbf{a} \cdot \mathbf{a} = a_1^2 + \cdots + a_n^2 = |\mathbf{a}|^2.$$

Let us show that the Cauchy-Schwartz Inequality (12.4) is also valid for vectors in Euclidean n-space. Since we cannot visualize the vectors geometrically when $n > 3$, it is necessary for us to give an algebraic proof. This proof is a bit complicated, and we give the details only for the case where $n = 2$. In this case

$$\mathbf{a} = [a_1, a_2], \qquad \mathbf{b} = [b_1, b_2], \qquad \mathbf{a} \cdot \mathbf{b} = a_1b_1 + a_2b_2,$$
$$|\mathbf{a}| = (a_1^2 + a_2^2)^{1/2}, \qquad |\mathbf{b}| = (b_1^2 + b_2^2)^{1/2}.$$

We are trying to prove that

$$|a_1b_1 + a_2b_2| \leqslant (a_1^2 + a_2^2)^{1/2}(b_1^2 + b_2^2)^{1/2}.$$

Both sides of the proposed inequality are nonnegative, and it therefore suffices to prove that

$$(a_1b_1 + a_2b_2)^2 \leq (a_1^2 + a_2^2)(b_1^2 + b_2^2),$$

that is,

$$a_1^2b_1^2 + 2a_1a_2b_1b_2 + a_2^2b_2^2 \leq a_1^2b_1^2 + a_1^2b_2^2 + a_2^2b_1^2 + a_2^2b_2^2.$$

Cancelling like terms and transposing, it is then sufficient to prove that

$$0 \leq a_1^2b_2^2 - 2a_1a_2b_1b_2 + a_2^2b_1^2,$$

that is,

(12.9) $$0 \leq (a_1b_2 - a_2b_1)^2.$$

But this is obviously true!

*[The proof for general n is analogous to that given above. In place of (12.9) we obtain

(12.10) $$\begin{aligned} 0 \leq (a_1b_2 - a_2b_1)^2 + (a_1b_3 - a_3b_1)^2 + \cdots + (a_1b_n - a_nb_1)^2 \\ + (a_2b_3 - a_3b_2)^2 + \cdots + (a_2b_n - a_nb_2)^2 \\ + \cdots + (a_{n-1}b_n - a_nb_{n-1})^2. \end{aligned}$$

Again this inequality holds, since the right-hand side is a sum of squares. Therefore the Cauchy-Schwartz Inequality (12.4) holds true for vectors in Euclidean n-space.]*

*[We may use formula (12.10) to decide when equality holds in the Cauchy-Schwartz Inequality, that is, when

(12.11) $$|\mathbf{a} \cdot \mathbf{b}| = |\mathbf{a}||\mathbf{b}|.$$

Of course (12.11) is valid if either **a** or **b** is the zero vector. We show

(12.12) Theorem

Suppose $\mathbf{a} \neq \mathbf{0}$. *Then equation* (12.11) *holds true if and only if* $\mathbf{b} = \lambda\mathbf{a}$ *for some scalar* λ.

PROOF: If $\mathbf{b} = \lambda\mathbf{a}$, then $|\mathbf{b}| = |\lambda||\mathbf{a}|$, and $|\mathbf{a}||\mathbf{b}| = |\lambda||\mathbf{a}|^2$. On the other hand,

$$|\mathbf{a} \cdot \mathbf{b}| = |\mathbf{a} \cdot (\lambda\mathbf{a})| = |\lambda(\mathbf{a} \cdot \mathbf{a})| = |\lambda||\mathbf{a} \cdot \mathbf{a}| = |\lambda||\mathbf{a}|^2.$$

Thus (12.11) is true when $\mathbf{b} = \lambda\mathbf{a}$.

Conversely, suppose (12.11) is valid. The method of proof of the Cauchy-Schwartz Inequality then shows that equality holds in (12.10). Therefore each of the expressions in parentheses on the right-hand side of (12.10) must be zero:

$$a_1b_2 - a_2b_1 = 0, \ \ldots, \ a_1b_n - a_nb_1 = 0, \ \ldots, \ a_{n-1}b_n - a_nb_{n-1} = 0.$$

Now $\mathbf{a} \neq \mathbf{0}$, so suppose for convenience that $a_1 \neq 0$. The above equations yield

$$b_1 = \lambda a_1, b_2 = \lambda a_2, \ldots, b_n = \lambda a_n, \qquad \lambda = b_1/a_1.$$

Therefore $\mathbf{b} = \lambda \mathbf{a}$, as claimed.]*

Let **a** and **b** be any pair of nonzero vectors. The Cauchy-Schwartz Inequality tells us that the scalar quantity

$$\frac{\mathbf{a} \cdot \mathbf{b}}{|\mathbf{a}||\mathbf{b}|}$$

lies between -1 and $+1$. We set this equal to $\cos \theta$, choosing θ so that $0° \leqslant \theta \leqslant 180°$, and then we *define* θ to be the *angle between the vectors* **a** and **b**. It follows from this definition that

$$\mathbf{a} \cdot \mathbf{b} = |\mathbf{a}||\mathbf{b}| \cos \theta.$$

*[If $\theta = 0°$, then $\mathbf{a} \cdot \mathbf{b} = |\mathbf{a}||\mathbf{b}|$. By (12.12) it follows that $\mathbf{b} = \lambda \mathbf{a}$ for some scalar λ. Then

$$\mathbf{a} \cdot \mathbf{b} = \mathbf{a} \cdot (\lambda \mathbf{a}) = \lambda(\mathbf{a} \cdot \mathbf{a}) = \lambda |\mathbf{a}|^2, \qquad |\mathbf{a}||\mathbf{b}| = |\lambda||\mathbf{a}|^2,$$

so $\lambda = |\lambda|$. This means that $\lambda > 0$. Hence if $\theta = 0°$, then $\mathbf{b} = \lambda \mathbf{a}$ for some positive scalar λ; the converse also holds, since the argument can be reversed. Likewise, $\theta = 180°$ if and only if $\mathbf{b} = \lambda \mathbf{a}$ for some $\lambda < 0$).]*

Another fundamental result is

(12.13) Triangle Inequality

For any vectors **a**, **b** *in Euclidean n-space,*

$$|\mathbf{a} + \mathbf{b}| \leqslant |\mathbf{a}| + |\mathbf{b}|.$$

PROOF: This is clear geometrically when $n = 3$: it is just the assertion that in Triangle OPQ, length of $OQ \leqslant$ length of OP + length of PQ.

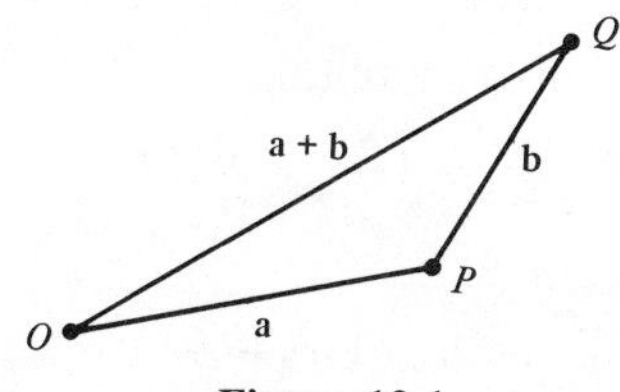

Figure 12.1

*[We may prove the inequality for vectors in n-space without much trouble. From (12.8) we obtain

$$\begin{aligned}|\mathbf{a}+\mathbf{b}|^2 &= (\mathbf{a}+\mathbf{b})\cdot(\mathbf{a}+\mathbf{b}) = \mathbf{a}\cdot\mathbf{a}+\mathbf{a}\cdot\mathbf{b}+\mathbf{b}\cdot\mathbf{a}+\mathbf{b}\cdot\mathbf{b}\\ &= |\mathbf{a}|^2+2\mathbf{a}\cdot\mathbf{b}+|\mathbf{b}|^2\\ &\leqslant |\mathbf{a}|^2+2|\mathbf{a}||\mathbf{b}|+|\mathbf{b}|^2. \qquad \text{(Cauchy-Schwartz!)}\end{aligned}$$

But

$$|\mathbf{a}|^2+2|\mathbf{a}||\mathbf{b}|+|\mathbf{b}|^2 = (|\mathbf{a}|+|\mathbf{b}|)^2,$$

so

$$|\mathbf{a}+\mathbf{b}|^2 \leqslant (|\mathbf{a}|+|\mathbf{b}|)^2.$$

Taking square roots, we get the desired inequality.]*

Let **a**, **b** be vectors in Euclidean n-space. Call them *orthogonal* (or *perpendicular*) if $\mathbf{a}\cdot\mathbf{b}=0$. If **a** and **b** are both nonzero, this means that $\cos\theta=0$, $\theta=90°$, where θ is the angle between **a** and **b**. Note that by definition **0** is orthogonal to every vector.

EXAMPLE

Let $\mathbf{a}=[1,0,-2,2]$, $\mathbf{b}=[5,-1,3,1]$. Then

$$|\mathbf{a}|=(1+4+4)^{1/2}=3, \qquad |\mathbf{b}|=(25+1+9+1)^{1/2}=6,$$

$$\mathbf{a}\cdot\mathbf{b}=1\cdot 5+(-2)\cdot 3+2\cdot 1=1,$$

$$\cos\theta=\frac{\mathbf{a}\cdot\mathbf{b}}{|\mathbf{a}||\mathbf{b}|}=\frac{1}{18}, \qquad \theta=\arccos\frac{1}{18}.$$

A unit vector in the direction of **a** is

$$\tfrac{1}{3}\mathbf{a}=\tfrac{1}{3}[1,0,-2,2]=[\tfrac{1}{3},0,\tfrac{-2}{3},\tfrac{2}{3}].$$

Now, continuing with this example, let us determine all vectors $\mathbf{c}=[c_1,c_2,c_3,c_4]$ orthogonal to both **a** and **b**. The conditions $\mathbf{a}\cdot\mathbf{c}=0$, $\mathbf{b}\cdot\mathbf{c}=0$ yield

$$c_1-2c_3+2c_4=0, \qquad 5c_1-c_2+3c_3+c_4=0.$$

Solving these simultaneously, we obtain

$$c_1=2c_3-2c_4, \qquad c_2=13c_3-9c_4, \qquad c_3,\ c_4 \text{ arbitrary.}$$

Therefore

$$\begin{aligned}\mathbf{c} &= [2c_3-2c_4,\ 13c_3-9c_4,\ c_3,c_4]\\ &= c_3[2,13,1,0]+c_4[-2,-9,0,1].\end{aligned}$$

Thus the set of all vectors **c** orthogonal to both **a** and **b** is a vector space, with basis $\{[2,13,1,0],\ [-2,-9,0,1]\}$.

The reader will recognize that we have solved the matrix equation

$$\begin{bmatrix} 1 & 0 & -2 & 2 \\ 5 & -1 & 3 & 1 \end{bmatrix} \begin{bmatrix} c_1 \\ c_2 \\ c_3 \\ c_4 \end{bmatrix} = \begin{bmatrix} 0 \\ 0 \end{bmatrix}.$$

The first matrix on the left has row vectors **a**, **b**, while the column vector on the left is $\mathbf{c}^T$ (the transpose of **c**). Then $\mathbf{c}^T$ ranges over all vectors in the null space of the matrix with rows **a**, **b**, while **c** ranges over all transposes of the vectors in that null space.

A set of vectors $\{\mathbf{u}_1, \ldots, \mathbf{u}_k\}$ in Euclidean n-space is *orthonormal* if each $\mathbf{u}_i$ is a unit vector, and each pair of vectors $\mathbf{u}_i$, $\mathbf{u}_j$ are orthogonal $(i \neq j)$. Here, "ortho" refers to "orthogonal," and "normal" to the fact that each $\mathbf{u}_i$ is a unit vector.

EXAMPLES

1. The vectors $\{\mathbf{i}, \mathbf{j}, \mathbf{k}\}$ form an orthonormal set in XYZ-space.
2. In 4-space, the vectors

$$[1,0,0,0], \left[0, \frac{-1}{\sqrt{2}}, \frac{1}{\sqrt{2}}, 0\right], \left[0, \frac{1}{\sqrt{2}}, \frac{1}{\sqrt{2}}, 0\right], [0,0,0,-1]$$

form an orthonormal set.

3. In 2-space, the vectors

$$[\cos\theta, -\sin\theta], [\sin\theta, \cos\theta]$$

form an orthonormal set.

An orthonormal set of vectors $\{\mathbf{u}_1, \ldots, \mathbf{u}_k\}$ in n-space may be represented geometrically by drawing k mutually perpendicular unit vectors, each starting at the origin. (Of course, we cannot really visualize this when $k > 3$.) It is intuitively clear that such a set of vectors must be linearly independent, that is, no one of them can be expressed as a linear combination of the others. In any case, we can prove this directly, as follows: suppose for instance that

$$\mathbf{u}_1 = c_2\mathbf{u}_2 + \cdots + c_k\mathbf{u}_k, \quad c_2, \ldots, c_k \text{ scalars.}$$

Then

$$\begin{aligned} 1 = \mathbf{u}_1 \cdot \mathbf{u}_1 &= \mathbf{u}_1 \cdot (c_2\mathbf{u}_2 + \cdots + c_k\mathbf{u}_k) \\ &= c_2(\mathbf{u}_1 \cdot \mathbf{u}_2) + \cdots + c_k(\mathbf{u}_1 \cdot \mathbf{u}_k) = 0, \end{aligned}$$

which is impossible.

We have thus shown that each orthonormal set of vectors $\{\mathbf{u}_1, \ldots, \mathbf{u}_k\}$ is a linearly independent set. These vectors therefore form a basis for the vector space spanned by them. We shall call $\{\mathbf{u}_1, \ldots, \mathbf{u}_k\}$ an *orthonormal basis* for this space.

Now let V be any vector space, with basis $\{\mathbf{v}_1, \ldots, \mathbf{v}_k\}$, where each $\mathbf{v}_i$ is a vector in Euclidean n-space. We shall show that V always has an orthonormal basis, and indeed we shall give the *Gram-Schmidt method* for obtaining such an orthonormal basis from the $\mathbf{v}$'s. To begin with, each $\mathbf{v}_i \neq \mathbf{0}$, since the $\mathbf{v}$'s form a basis. Normalize $\mathbf{v}_1$ (that is, find a unit vector $\mathbf{u}_1$ in the same direction as $\mathbf{v}_1$) by setting

$$\mathbf{u}_1 = \frac{1}{|\mathbf{v}_1|}\,\mathbf{v}_1.$$

For each scalar λ, the vector $\mathbf{v}_2' = \mathbf{v}_2 - \lambda\mathbf{u}_1$ lies in V. Let us determine λ so that the vector $\mathbf{v}_2'$ is orthogonal to $\mathbf{u}_1$. Indeed,

$$\mathbf{u}_1 \cdot \mathbf{v}_2' = \mathbf{u}_1 \cdot (\mathbf{v}_2 - \lambda\mathbf{u}_1) = \mathbf{u}_1 \cdot \mathbf{v}_2 - \lambda(\mathbf{u}_1 \cdot \mathbf{u}_1) = \mathbf{u}_1 \cdot \mathbf{v}_2 - \lambda,$$

so we need only pick $\lambda = \mathbf{u}_1 \cdot \mathbf{v}_2$. Thus, we set

$$\mathbf{v}_2' = \mathbf{v}_2 - (\mathbf{u}_1 \cdot \mathbf{v}_2)\mathbf{u}_1.$$

Now normalize $\mathbf{v}_2'$, getting a unit vector $\mathbf{u}_2$ which is also orthogonal to $\mathbf{u}_1$.

Continuing the process, we set

$$\mathbf{v}_3' = \mathbf{v}_3 - (\mathbf{u}_1 \cdot \mathbf{v}_3)\mathbf{u}_1 - (\mathbf{u}_2 \cdot \mathbf{v}_3)\mathbf{u}_2,$$

a vector in V orthogonal to both $\mathbf{u}_1$ and $\mathbf{u}_2$. Normalize $\mathbf{v}_3'$ to obtain a unit vector $\mathbf{u}_3$ orthogonal to $\mathbf{u}_1$ and $\mathbf{u}_2$. For the next step,

$$\mathbf{v}_4' = \mathbf{v}_4 - (\mathbf{u}_1 \cdot \mathbf{v}_4)\mathbf{u}_1 - (\mathbf{u}_2 \cdot \mathbf{v}_4)\mathbf{u}_2 - (\mathbf{u}_3 \cdot \mathbf{v}_4)\mathbf{u}_3,$$

and so on.

In this manner we obtain an orthonormal set of vectors $\{\mathbf{u}_1, \ldots, \mathbf{u}_k\}$ all lying in V.[3] Since we can solve for $\mathbf{v}_1, \mathbf{v}_2, \ldots, \mathbf{v}_k$ in terms of the $\mathbf{u}$'s, every linear combination of the $\mathbf{v}$'s is also a linear combination of the $\mathbf{u}$'s. Hence the set of vectors $\{\mathbf{u}_1, \ldots, \mathbf{u}_k\}$ spans V, and is therefore an orthonormal basis for V.

[3] In order to be sure that the procedure works, we must know that each of the vectors $\mathbf{v}_2', \ldots, \mathbf{v}_k'$ is not zero, since otherwise we could not normalize them. If $\mathbf{v}_2' = \mathbf{0}$, then $\mathbf{v}_2$ would be a scalar multiple of $\mathbf{u}_1$, hence also of $\mathbf{v}_1$; this is impossible, since $\{\mathbf{v}_1, \mathbf{v}_2, \ldots, \mathbf{v}_k\}$ is a linearly independent set. Likewise, if $\mathbf{v}_3' = \mathbf{0}$, then $\mathbf{v}_3$ would be a linear combination of $\mathbf{u}_1$ and $\mathbf{u}_2$, hence of $\mathbf{v}_1$ and $\mathbf{v}_2$, which is again impossible. Continuing in this way, we see that none of the vectors $\mathbf{v}_2', \ldots, \mathbf{v}_k'$ is $\mathbf{0}$.

EXAMPLE

Let V be the vector space with basis

$$\mathbf{v}_1 = [-1,0,0,0], \qquad \mathbf{v}_2 = [1,0,2,2], \qquad \mathbf{v}_3 = [0,0,3,4].$$

We shall use the Gram-Schmidt method to obtain an orthonormal basis for V. We have

$$|\mathbf{v}_1| = 1, \qquad \mathbf{u}_1 = \mathbf{v}_1, \qquad \mathbf{u}_1 \cdot \mathbf{v}_2 = -1, \qquad \mathbf{u}_1 \cdot \mathbf{v}_3 = 0.$$

Then

$$\mathbf{v}_2' = \mathbf{v}_2 - (\mathbf{u}_1 \cdot \mathbf{v}_2)\mathbf{u}_1 = [1,0,2,2] + [-1,0,0,0] = [0,0,2,2],$$

$$\mathbf{u}_2 = \frac{1}{|\mathbf{v}_2'|}\mathbf{v}_2' = \frac{1}{2\sqrt{2}}\mathbf{v}_2' = \frac{1}{\sqrt{2}}[0,0,1,1], \qquad \mathbf{u}_2 \cdot \mathbf{v}_3 = \frac{7}{\sqrt{2}}.$$

Next,

$$\begin{aligned}\mathbf{v}_3' &= \mathbf{v}_3 - (\mathbf{u}_1 \cdot \mathbf{v}_3)\mathbf{u}_1 - (\mathbf{u}_2 \cdot \mathbf{v}_3)\mathbf{u}_2 \\ &= [0,0,3,4] - \frac{7}{\sqrt{2}} \cdot \frac{1}{\sqrt{2}}[0,0,1,1] = [0,0,-\tfrac{1}{2},\tfrac{1}{2}].\end{aligned}$$

Then

$$|\mathbf{v}_3'| = \frac{\sqrt{2}}{2}, \qquad \mathbf{u}_3 = \frac{2}{\sqrt{2}}\mathbf{v}_3' = \left[0,0,\frac{-1}{\sqrt{2}},\frac{1}{\sqrt{2}}\right].$$

The vectors $\{\mathbf{u}_1, \mathbf{u}_2, \mathbf{u}_3\}$ form an orthonormal basis for V, where

$$\mathbf{u}_1 = [-1,0,0,0], \qquad \mathbf{u}_2 = \left[0,0,\frac{1}{\sqrt{2}},\frac{1}{\sqrt{2}}\right], \qquad \mathbf{u}_3 = \left[0,0,\frac{-1}{\sqrt{2}},\frac{1}{\sqrt{2}}\right].$$

EXERCISES

1. Let $\mathbf{a} = [1, 2, -2, 0]$, $\quad \mathbf{b} = [2, 1, 0, -2]$.
   (a) Find the lengths of $\mathbf{a}$, $\mathbf{b}$, and $\mathbf{a} + \mathbf{b}$. What is the angle between $\mathbf{a}$ and $\mathbf{b}$?
   (b) Find all vectors $\mathbf{c}$ orthogonal to both $\mathbf{a}$ and $\mathbf{b}$.
   (c) Find an orthonormal basis for the vector space spanned by $\mathbf{a}$ and $\mathbf{b}$.
   (d) Find an orthonormal basis for the set of vectors $\mathbf{c}$ in part (b).
2. Prove that every unit vector in Euclidean 2-space is of the form $[\cos\theta, \sin\theta]$ for some θ.

3. Prove that $[1,0,0,0]$, $[0,1,0,0]$, $[0,0,1,0]$, $[0,0,0,1]$ is an orthonormal basis for 4-space. Are there any other orthonormal bases?
4. Use the Gram-Schmidt method to find an orthonormal basis for the vector space spanned by $[0,1,0]$, $[3,-1,4]$, $[2,2,-1]$.
5. Let $\{\mathbf{u}_1, \ldots, \mathbf{u}_k\}$ be an orthonormal set of vectors, and let $\alpha_1, \ldots, \alpha_k$ be scalars. Prove that

$$|\alpha_1\mathbf{u}_1 + \cdots + \alpha_k\mathbf{u}_k| = (\alpha_1^2 + \cdots + \alpha_k^2)^{1/2}.$$

[*Hint:* The square of the left-hand expression is

$$(\alpha_1\mathbf{u}_1 + \cdots + \alpha_k\mathbf{u}_k) \cdot (\alpha_1\mathbf{u}_1 + \cdots + \alpha_k\mathbf{u}_k).]$$

6. Let $\{\mathbf{u}_1, \ldots, \mathbf{u}_k\}$ be an orthonormal basis for V. Show that each vector $\mathbf{v}$ in V can be expressed as

$$\mathbf{v} = \alpha_1\mathbf{u}_1 + \cdots + \alpha_k\mathbf{u}_k, \quad \text{where } \alpha_i = \mathbf{u}_i \cdot \mathbf{v}, \qquad 1 \leqslant i \leqslant k.$$

7. Show that an orthonormal set of vectors $\{\mathbf{u}_1, \ldots, \mathbf{u}_k\}$ in Euclidean n-space cannot contain more than n vectors. [*Hint:* By Exercise 9.11, a linearly independent set of vectors in n-space cannot consist of more than n vectors.]
8. Let V be the vector space spanned by $\{\mathbf{v}_1, \ldots, \mathbf{v}_k\}$, with each $\mathbf{v}_i$ a vector in Euclidean n-space. Let $\mathbf{w}$ be a vector in n-space orthogonal to each $\mathbf{v}_i$. Show that $\mathbf{w}$ is orthogonal to every vector in V.

*9. Show that if $\{\mathbf{a}, \mathbf{b}\}$ is an orthonormal basis for Euclidean 2-space, then either

$$\mathbf{a} = [\cos\theta, \sin\theta], \qquad \mathbf{b} = [-\sin\theta, \cos\theta] \quad \text{for some } \theta,$$

or else

$$\mathbf{a} = [\cos\theta, \sin\theta], \qquad \mathbf{b} = [\sin\theta, -\cos\theta] \quad \text{for some } \theta.$$

10. Show that every unit vector in Euclidean 3-space is of the form $[\cos\theta \sin\varphi, \sin\theta \sin\varphi, \cos\varphi]$ for some angles θ and φ. [*Hint:* Use spherical coordinates; see Exercise 15.2.]

*11. Let $\mathbf{a}$, $\mathbf{b}$ be nonzero vectors in Euclidean n-space. Prove that

$$|\mathbf{a} + \mathbf{b}| = |\mathbf{a}| + |\mathbf{b}|$$

is true if and only if there is a positive scalar λ such that $\mathbf{b} = \lambda\mathbf{a}$. [*Hint:* Show that the displayed equality holds true if and only if $\mathbf{a} \cdot \mathbf{b} = |\mathbf{a}||\mathbf{b}|$, and then use Theorem 12.12.]

13 ORTHOGONAL MATRICES

An *orthogonal matrix* is a square matrix A such that

(13.1) $$A \cdot A^{\mathrm{T}} = I.$$

If this equation holds, then A must be nonsingular, and A^{T} must equal A^{-1}. Therefore also $A^{\mathrm{T}} \cdot A = I$. The argument may be reversed: if $A^{\mathrm{T}} \cdot A = I$ then also $A \cdot A^{\mathrm{T}} = I$, and A is orthogonal.

EXAMPLES

Each of the following matrices is orthogonal:

$$\begin{bmatrix} -1 & 0 \\ 0 & 1 \end{bmatrix}, \begin{bmatrix} \cos\theta & -\sin\theta \\ \sin\theta & \cos\theta \end{bmatrix}, \begin{bmatrix} \cos\theta & \sin\theta \\ \sin\theta & -\cos\theta \end{bmatrix}, \begin{bmatrix} \frac{3}{5} & \frac{-4}{5} & 0 \\ \frac{4}{5} & \frac{3}{5} & 0 \\ 0 & 0 & 1 \end{bmatrix}.$$

Let us recall some definitions from section 12. Let **a**, **b** be a pair of vectors in Euclidean n-space, and suppose that

$$\mathbf{a} = [a_1 \quad \cdots \quad a_n], \qquad \mathbf{b} = [b_1 \quad \cdots \quad b_n].$$

The *dot product* $\mathbf{a} \cdot \mathbf{b}$ is defined as

$$\mathbf{a} \cdot \mathbf{b} = a_1 b_1 + \cdots + a_n b_n.$$

If $\mathbf{a} \cdot \mathbf{a} = 1$, call **a** a *unit* vector. An *orthonormal set* of vectors in Euclidean n-space is a set of unit vectors $\{\mathbf{u}_1, \ldots, \mathbf{u}_k\}$ in that space, such that $\mathbf{u}_i \cdot \mathbf{u}_j = 0$ whenever $i \neq j$. For example, let

$$\mathbf{u}_1 = [\cos\theta, -\sin\theta], \qquad \mathbf{u}_2 = [\sin\theta, \cos\theta].$$

Then $\{\mathbf{u}_1, \mathbf{u}_2\}$ is an orthonormal set of vectors in Euclidean 2-space. Notice that the following is an orthogonal matrix:

$$\begin{bmatrix} \cos\theta & -\sin\theta \\ \sin\theta & \cos\theta \end{bmatrix} = \begin{bmatrix} \mathbf{u}_1 \\ \mathbf{u}_2 \end{bmatrix}.$$

There is always such a relation between orthogonal matrices and orthonormal sets of vectors:

(13.2) Theorem

A square matrix is orthogonal if and only if its row vectors are an orthonormal set of vectors.

*[*PROOF:* For simplicity we give the proof for the case of 3×3 matrices. Let A be any 3×3 matrix with row vectors $\mathbf{r}_1$, $\mathbf{r}_2$, $\mathbf{r}_3$, and let

$$\mathbf{r}_1 = [a_1 \quad a_2 \quad a_3], \qquad \mathbf{r}_2 = [b_1 \quad b_2 \quad b_3], \qquad \mathbf{r}_3 = [c_1 \quad c_2 \quad c_3].$$

Then

$$A \cdot A^{\mathrm{T}} = \begin{bmatrix} a_1 & a_2 & a_3 \\ b_1 & b_2 & b_3 \\ c_1 & c_2 & c_3 \end{bmatrix} \begin{bmatrix} a_1 & b_1 & c_1 \\ a_2 & b_2 & c_2 \\ a_3 & b_3 & c_3 \end{bmatrix}$$

$$= \begin{bmatrix} a_1a_1 + a_2a_2 + a_3a_3 & a_1b_1 + a_2b_2 + a_3b_3 & a_1c_1 + a_2c_2 + a_3c_3 \\ b_1a_1 + b_2a_2 + b_3a_3 & b_1b_1 + b_2b_2 + b_3b_3 & b_1c_1 + b_2c_2 + b_3c_3 \\ c_1a_1 + c_2a_2 + c_3a_3 & c_1b_1 + c_2b_2 + c_3b_3 & c_1c_1 + c_2c_2 + c_3c_3 \end{bmatrix}$$

$$= \begin{bmatrix} \mathbf{r}_1 \cdot \mathbf{r}_1 & \mathbf{r}_1 \cdot \mathbf{r}_2 & \mathbf{r}_1 \cdot \mathbf{r}_3 \\ \mathbf{r}_2 \cdot \mathbf{r}_1 & \mathbf{r}_2 \cdot \mathbf{r}_2 & \mathbf{r}_2 \cdot \mathbf{r}_3 \\ \mathbf{r}_3 \cdot \mathbf{r}_1 & \mathbf{r}_3 \cdot \mathbf{r}_2 & \mathbf{r}_3 \cdot \mathbf{r}_3 \end{bmatrix}.$$

But A is orthogonal if and only if $AA^{\mathrm{T}} = I$, that is, if and only if the above matrix is the identity matrix. Hence A is orthogonal if and only if the following equations hold true:

$$\mathbf{r}_1 \cdot \mathbf{r}_1 = \mathbf{r}_2 \cdot \mathbf{r}_2 = \mathbf{r}_3 \cdot \mathbf{r}_3 = 1, \qquad \mathbf{r}_1 \cdot \mathbf{r}_2 = \mathbf{r}_2 \cdot \mathbf{r}_3 = \mathbf{r}_3 \cdot \mathbf{r}_1 = 0.$$

However, these are precisely the conditions that $\{\mathbf{r}_1, \mathbf{r}_2, \mathbf{r}_3\}$ be an orthonormal set of vectors. This shows that A is an orthogonal matrix if and only if its rows form an orthonormal set of vectors. An analogous proof holds for $n \times n$ matrices.]*

The same type of reasoning also shows

(13.3) Theorem

A square matrix is orthogonal if and only if its column vectors are an orthonormal set of vectors.

Now let A be any orthogonal $n \times n$ matrix. Then $AA^{\mathrm{T}} = I$, so

$$(\det A)(\det A^{\mathrm{T}}) = 1.$$

By Theorem 5.5, part (v), $\det A^{\mathrm{T}} = \det A$. Therefore $(\det A)^2 = 1$, so $\det A$ is either $+1$ or -1. Call A a *proper* orthogonal matrix if $\det A = +1$, and an *improper* orthogonal matrix if $\det A = -1$.

We shall next discuss the effect of a linear transformation on Euclidean n-space, in the case where the matrix which determines the transformation is a real orthogonal matrix. Let $\mathbf{v}$ denote an arbitrary $n \times 1$ column vector with real entries, and let A be an $n \times n$ matrix whose entries are also real numbers. The mapping, which assigns to each vector $\mathbf{v}$ the vector $A\mathbf{v}$, is called a *linear transformation* of Euclidean n-space into itself. When A is an orthogonal matrix, we call the mapping an *orthogonal transformation.*

We are now going to prove that every orthogonal transformation preserves lengths of vectors and also preserves angles between pairs of vectors. To be specific, we shall show

(13.4) Theorem

Let $A^{n \times n}$ be a real orthogonal matrix. Then for every $n \times 1$ column vector $\mathbf{v}$ with real entries,

$$\text{length of } \mathbf{v} = \text{length of } A\mathbf{v}.$$

Furthermore, for each pair of nonzero real vectors $\mathbf{v}$ and $\mathbf{w}$,

$$\text{angle between } \mathbf{v} \text{ and } \mathbf{w} = \text{angle between } A\mathbf{v} \text{ and } A\mathbf{w}.$$

*[*PROOF:* Let $\mathbf{v}$ and $\mathbf{w}$ be any $n \times 1$ column vectors with real entries, say

$$\mathbf{v} = \begin{bmatrix} x_1 \\ \cdot \\ \cdot \\ \cdot \\ x_n \end{bmatrix}, \qquad \mathbf{w} = \begin{bmatrix} y_1 \\ \cdot \\ \cdot \\ \cdot \\ y_n \end{bmatrix}.$$

Then by definition

$$\mathbf{v} \cdot \mathbf{w} = x_1 y_1 + \cdots + x_n y_n.$$

On the other hand

$$\textbf{(13.5)} \qquad \mathbf{v}^{\mathrm{T}}\,\mathbf{w} = [x_1 \quad \cdots \quad x_n] \begin{bmatrix} y_1 \\ \cdot \\ \cdot \\ \cdot \\ y_n \end{bmatrix} = x_1 y_1 + \cdots + x_n y_n.$$

(In the last equation, we have dropped the distinction between a scalar α and the 1×1 matrix $[\alpha]$.) Comparing the above two formulas, we obtain

$$\textbf{(13.6)} \qquad \mathbf{v} \cdot \mathbf{w} = \mathbf{v}^{\mathrm{T}}\,\mathbf{w}.$$

This shows that we can compute the dot product $\mathbf{v} \cdot \mathbf{w}$ of the vectors $\mathbf{v}$ and $\mathbf{w}$ by computing the matrix product $\mathbf{v}^{\mathrm{T}}\,\mathbf{w}$ of the $1 \times n$ matrix $\mathbf{v}^{\mathrm{T}}$ with the $n \times 1$ matrix $\mathbf{w}$.

Now let A be any orthogonal $n \times n$ matrix. Applying (13.6) with $A\mathbf{v}$ in place of $\mathbf{v}$, and $A\mathbf{w}$ in place of $\mathbf{w}$, we get

$$\begin{aligned}(A\mathbf{v}) \cdot (A\mathbf{w}) &= (A\mathbf{v})^{\mathrm{T}}(A\mathbf{w}) = (\mathbf{v}^{\mathrm{T}}A^{\mathrm{T}})(A\mathbf{w}) \\ &= \mathbf{v}^{\mathrm{T}}(A^{\mathrm{T}}A)\mathbf{w} = \mathbf{v}^{\mathrm{T}}\ I\ \mathbf{w} = \mathbf{v}^{\mathrm{T}}\ \mathbf{w} = \mathbf{v} \cdot \mathbf{w}.\end{aligned}$$

This proves that an orthogonal transformation preserves dot products of vectors. Taking the case where $\mathbf{w} = \mathbf{v}$, we obtain

$$|\mathbf{v}| = (\mathbf{v} \cdot \mathbf{v})^{1/2} = (A\mathbf{v} \cdot A\mathbf{v})^{1/2} = |A\mathbf{v}|.$$

Likewise, if $\mathbf{v}$ and $\mathbf{w}$ are nonzero vectors, and ϕ is the angle between them, then

$$\cos \phi = \frac{\mathbf{v} \cdot \mathbf{w}}{|\mathbf{v}||\mathbf{w}|} = \frac{A\mathbf{v} \cdot A\mathbf{w}}{|A\mathbf{v}||A\mathbf{w}|}.$$

But the last expression is the cosine of the angle between $A\mathbf{v}$ and $A\mathbf{w}$, and so that angle equals ϕ. This completes the proof.]*

Remarks:

1. Although an orthogonal transformation preserves angles between vectors, it may possibly reverse the direction of the rotation which carries one vector onto the other. This is discussed in more detail below.
2. There is a converse of Theorem 13.4 which also holds true: *any linear transformation which preserves lengths of vectors must be an orthogonal transformation.* This result is somewhat harder to prove.[1]

We are now ready to study the effect of an orthogonal transformation on the XY-plane. At the same time, our discussion will enable us to find all real orthogonal 2×2 matrices.

(13.7) Theorem

Let A be a real orthogonal 2×2 matrix. If det $A = +1$, then

(13.8)
$$A = \begin{bmatrix} \cos\theta & -\sin\theta \\ \sin\theta & \cos\theta \end{bmatrix}$$

for some angle θ. The orthogonal transformation $\mathbf{v} \rightarrow A\mathbf{v}$ is a rotation through the angle θ about the origin.

If det $A = -1$, then

(13.9)
$$A = \begin{bmatrix} \cos\theta & \sin\theta \\ \sin\theta & -\cos\theta \end{bmatrix}$$

for some angle θ. In this case, the orthogonal transformation $\mathbf{v} \rightarrow A\mathbf{v}$ is

[1] See references in the Preface for the proof.

a rotation through θ about the origin, followed by a reflection across the line through the origin with angle of inclination θ.

Conversely, for each angle θ, the matrices in (13.8) *and* (13.9) *are real orthogonal matrices with determinant* $+1, -1$, *respectively.*

*[*PROOF:* Let P denote an arbitrary point in the XY-plane, and let (x,y) be its coordinates. Set $\mathbf{v} = \overrightarrow{OP}$, and represent $\mathbf{v}$ as a column vector $\begin{bmatrix} x \\ y \end{bmatrix}$. Let $\mathbf{i} = \begin{bmatrix} 1 \\ 0 \end{bmatrix}$, $\mathbf{j} = \begin{bmatrix} 0 \\ 1 \end{bmatrix}$, so that $\mathbf{i}$, $\mathbf{j}$ are the usual unit vectors along the coordinate axes in the XY-plane. We may write

$$\mathbf{v} = \begin{bmatrix} x \\ y \end{bmatrix} = x \begin{bmatrix} 1 \\ 0 \end{bmatrix} + y \begin{bmatrix} 0 \\ 1 \end{bmatrix} = x\mathbf{i} + y\mathbf{j}.$$

The linear transformation determined by a 2×2 matrix A carries each vector $\mathbf{v}$ onto a vector $\mathbf{v}'$, where $\mathbf{v}' = A\mathbf{v}$. If we write $\mathbf{v}' = \overrightarrow{OP'} = \begin{bmatrix} x' \\ y' \end{bmatrix}$, then (by definition)

$$\begin{bmatrix} x' \\ y' \end{bmatrix} = A \begin{bmatrix} x \\ y \end{bmatrix}.$$

Now suppose that A is a real orthogonal matrix. Since $\mathbf{i}$, $\mathbf{j}$ are mutually perpendicular unit vectors, it follows from Theorem 13.4 that $A\mathbf{i}$ and $A\mathbf{j}$ are also mutually perpendicular unit vectors. Let us set

$$\mathbf{u}_1 = \overrightarrow{OP_1} = A\mathbf{i}, \qquad \mathbf{u}_2 = \overrightarrow{OP_2} = A\mathbf{j}.$$

The points P_1 and P_2 lie on a unit circle with center at O, since $\mathbf{u}_1$ and $\mathbf{u}_2$ are unit vectors.

Let (x_1,y_1) be the coordinates of P_1, and (x_2,y_2) those of P_2; then $\mathbf{u}_1 = \begin{bmatrix} x_1 \\ y_1 \end{bmatrix}$, $\mathbf{u}_2 = \begin{bmatrix} x_2 \\ y_2 \end{bmatrix}$. However, we also have

$$\mathbf{u}_1 = A\mathbf{i} = A \begin{bmatrix} 1 \\ 0 \end{bmatrix} = \text{1st column of } A,$$

$$\mathbf{u}_2 = A\mathbf{j} = A \begin{bmatrix} 0 \\ 1 \end{bmatrix} = \text{2nd column of } A.$$

Therefore

(13.10)
$$A = \begin{bmatrix} x_1 & x_2 \\ y_1 & y_2 \end{bmatrix}.$$

Suppose now that θ denotes the angle of inclination of $\overrightarrow{OP_1}$, that is, the angle through which OX must be rotated to reach the direction $\overrightarrow{OP_1}$. Then of course

$$x_1 = \cos\theta, \qquad y_1 = \sin\theta.$$

In order to find the coordinates (x_2, y_2) of P_2, we shall have to consider two cases (Figure 13.1(a) and (b)) depending on the position of $\mathbf{u}_2$ relative to $\mathbf{u}_1$. (In either case, $\mathbf{u}_2$ must be perpendicular to $\mathbf{u}_1$.)

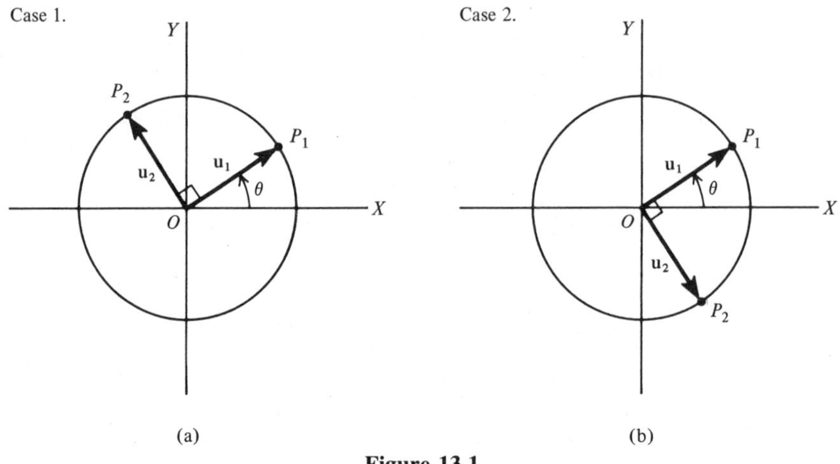

Figure 13.1

CASE 1

Suppose that a counterclockwise 90° rotation carries $\mathbf{u}_1$ onto $\mathbf{u}_2$. Then

$$x_2 = \cos(\theta + 90°) = \cos\theta \cos 90° - \sin\theta \sin 90° = -\sin\theta,$$
$$y_2 = \sin(\theta + 90°) = \sin\theta \cos 90° + \cos\theta \sin 90° = \cos\theta.$$

Substituting into (13.10), we obtain

$$A = \begin{bmatrix} \cos\theta & -\sin\theta \\ \sin\theta & \cos\theta \end{bmatrix}.$$

Note that $\det A = +1$ in this case.

Again let P denote an arbitrary point in the XY-plane, so that as pointed out earlier,

$$\mathbf{v} = x\mathbf{i} + y\mathbf{j}.$$

Then

$$\mathbf{v}' = A\mathbf{v} = A(x\mathbf{i} + y\mathbf{j}) = x(A\mathbf{i}) + y(A\mathbf{j}) = x\mathbf{u}_1 + y\mathbf{u}_2.$$

This shows that $\mathbf{v}'$ has the same relation to the pair of vectors $\mathbf{u}_1$, $\mathbf{u}_2$ that $\mathbf{v}$ has to the pair $\mathbf{i}$, $\mathbf{j}$. But $\mathbf{u}_1$, $\mathbf{u}_2$ are gotten by rotating $\mathbf{i}$, $\mathbf{j}$, respectively, through the angle θ. Therefore $\mathbf{v}'$ is obtained by rotating $\mathbf{v}$ through the angle θ. This proves that the orthogonal transformation corresponding to the matrix (13.8) is a rotation through θ about the origin.

This is illustrated geometrically in Figure 13.2.

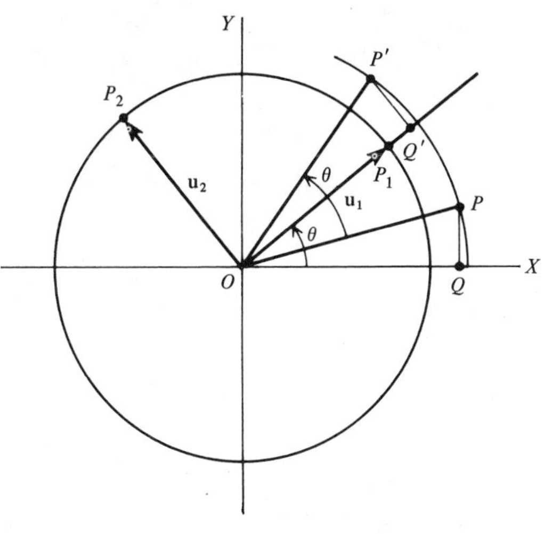

Figure 13.2

In this diagram,

$$\theta = \measuredangle XOP_1 = \measuredangle POP',$$
$$\overrightarrow{OP} = \overrightarrow{OQ} + \overrightarrow{QP} = x\mathbf{i} + y\mathbf{j}, \qquad \overrightarrow{OQ} = x\mathbf{i}, \qquad \overrightarrow{QP} = y\mathbf{j},$$
$$\overrightarrow{OP'} = \overrightarrow{OQ'} + \overrightarrow{Q'P'} = x\mathbf{u}_1 + y\mathbf{u}_2, \qquad \overrightarrow{OQ'} = x\mathbf{u}_1, \qquad \overrightarrow{Q'P'} = y\mathbf{u}_2.$$

CASE 2

Suppose that a *clockwise* 90° rotation carries $\mathbf{u}_1$ onto $\mathbf{u}_2$. In this case

$$x_2 = \cos(\theta - 90°) = \sin\theta, \qquad y_2 = \sin(\theta - 90°) = -\cos\theta.$$

Therefore by (13.10),

$$A = \begin{bmatrix} \cos\theta & \sin\theta \\ \sin\theta & -\cos\theta \end{bmatrix}.$$

Note that det $A = -1$ this time.

Let P be an arbitrary point in the XY-plane. Again we have

$$\mathbf{v} = x\mathbf{i} + y\mathbf{j}, \qquad \mathbf{v}' = A\mathbf{v} = x\mathbf{u}_1 + y\mathbf{u}_2.$$

Our picture now looks as follows:

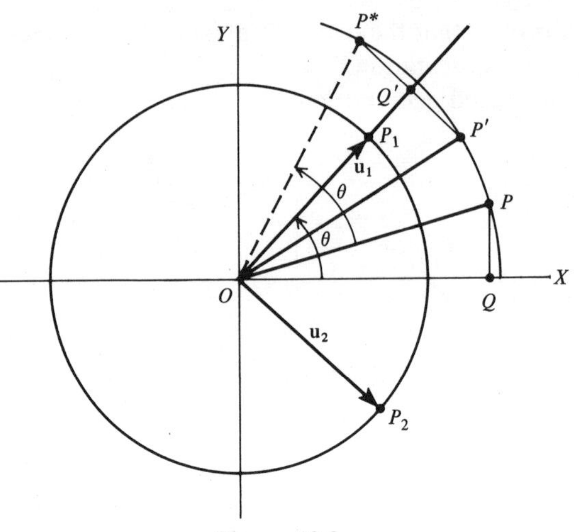

Figure 13.3

In Figure 13.3,

$$\theta = \measuredangle XOP_1 = \measuredangle POP^*,$$

$$\overrightarrow{OP} = \overrightarrow{OQ} + \overrightarrow{QP} = x\mathbf{i} + y\mathbf{j}, \qquad \overrightarrow{OQ} = x\mathbf{i}, \qquad \overrightarrow{QP} = y\mathbf{j},$$

$$\overrightarrow{OP'} = \overrightarrow{OQ'} + \overrightarrow{Q'P'} = x\mathbf{u}_1 + y\mathbf{u}_2, \qquad \overrightarrow{OQ'} = x\mathbf{u}_1, \qquad \overrightarrow{Q'P'} = y\mathbf{u}_2.$$

It is clear geometrically that P' is the reflection of P^* across the line OQ'. Thus we may obtain $\overrightarrow{OP'}$ from $\overrightarrow{OP}$ by first rotating $\overrightarrow{OP}$ through an angle θ to the position $\overrightarrow{OP^*}$, and then reflecting $\overrightarrow{OP^*}$ across the line $\overrightarrow{OQ'}$ to obtain the vector $\overrightarrow{OP'}$.

We have now shown that every real orthogonal 2×2 matrix A is given by either (13.8) or (13.9), and we have given the geometric significance of the corresponding orthogonal transformation. Alternative (13.8) occurs only when $\det A = +1$, and (13.9) only when $\det A = -1$.

It is simple to verify that, conversely, for each angle θ the formulas (13.8) and (13.9) do indeed give real orthogonal matrices of determinants $+1, -1$, respectively. We leave this verification to the reader. This completes the proof of Theorem 13.7.]*

Remarks:

1. The improper orthogonal transformation determined by the orthogonal matrix

$$\begin{bmatrix} \cos\theta & \sin\theta \\ \sin\theta & -\cos\theta \end{bmatrix}$$

can also be described as follows: it maps an arbitrary vector $\overrightarrow{OP}$ onto the vector $\overrightarrow{OP'}$, obtained by reflecting $\overrightarrow{OP}$ across the line through the origin with angle of inclination $\theta/2$. This can be verified by a geometrical argument; it also follows as a consequence of Exercise 8.9.

2. It is harder to describe all orthogonal 3×3 matrices. Nevertheless, we may still give a geometric interpretation of the orthogonal transformation $\mathbf{v} \to A\mathbf{v}$ determined by a real orthogonal 3×3 matrix A. Specifically when $\det A = +1$, this orthogonal transformation is given by a rotation about some axis L passing through the origin.[2] On the other hand, when $\det A = -1$, the orthogonal transformation $\mathbf{v} \to A\mathbf{v}$ is given by such a rotation, followed by a reflection across the plane through O perpendicular to the axis of rotation L.

3. More generally, let A be a real $n \times n$ orthogonal matrix, and consider the orthogonal transformation $\mathbf{v} \to A\mathbf{v}$ of Euclidean n-space into itself. If $\det A = +1$, the transformation is given geometrically by a succession of several rotations about different axes through the origin O. When $\det A = -1$, the transformation is a succession of such rotations, followed by a reflection.[3]

EXERCISES

1. Find all diagonal real orthogonal $n \times n$ matrices.
2. Find an orthogonal 2×2 matrix A for which the orthogonal transformation $\mathbf{v} \to A\mathbf{v}$ of Euclidean 2-space is given thus:
   (a) a counterclockwise rotation through 30° about the origin.
   (b) a reflection across the Y-axis.
   (c) a clockwise rotation through θ about the origin.
3. Let

$$A = \begin{bmatrix} \cos\theta & -\sin\theta & 0 \\ \sin\theta & \cos\theta & 0 \\ 0 & 0 & 1 \end{bmatrix}, \qquad B = \begin{bmatrix} \cos\theta & -\sin\theta & 0 \\ \sin\theta & \cos\theta & 0 \\ 0 & 0 & -1 \end{bmatrix}.$$

   Show that A is a proper orthogonal matrix and B an improper orthogonal matrix.

*4. Let A be as in Exercise 3. Show that the orthogonal transformation $\begin{bmatrix} x \\ y \\ z \end{bmatrix} \longrightarrow A \begin{bmatrix} x \\ y \\ z \end{bmatrix}$ is given geometrically by a rotation through the angle θ about OZ.

[2] See references in the Preface for the proof.
[3] See references in the Preface for proofs.

*5. Given the matrix B in Exercise 3, show that the orthogonal transformation $\begin{bmatrix} x \\ y \\ z \end{bmatrix} \longrightarrow B \begin{bmatrix} x \\ y \\ z \end{bmatrix}$ is given by a rotation through θ about OZ, followed by a reflection in the XY-plane.

6. Find all possible values of x and y for which $\begin{bmatrix} 1 & x \\ 0 & y \end{bmatrix}$ is an orthogonal matrix.

7. Find all possible values of x, y, z such that the matrix

$$\begin{bmatrix} \frac{1}{\sqrt{2}} & \frac{2}{3} & x \\ \frac{1}{\sqrt{2}} & \frac{-2}{3} & y \\ 0 & \frac{1}{3} & z \end{bmatrix}$$

is orthogonal.

8. For the orthogonal matrix A given by (13.8), what is A^{-1}? Describe the orthogonal transformation $\mathbf{v} \rightarrow A^{-1}\mathbf{v}$ geometrically.

9. Let

$$A = \begin{bmatrix} \cos\alpha & -\sin\alpha \\ \sin\alpha & \cos\alpha \end{bmatrix}, \qquad B = \begin{bmatrix} \cos\beta & -\sin\beta \\ \sin\beta & \cos\beta \end{bmatrix}.$$

Show that

$$AB = \begin{bmatrix} \cos(\alpha+\beta) & -\sin(\alpha+\beta) \\ \sin(\alpha+\beta) & \cos(\alpha+\beta) \end{bmatrix}.$$

Could you have predicted this result from Theorem 13.7?

10. Show that if A is orthogonal, so is A^{T}. [*Hint:* Take transposes of both sides of the equation $AA^{\mathrm{T}} = I$, thereby obtaining $(A^{\mathrm{T}})^{\mathrm{T}}A^{\mathrm{T}} = I$. Therefore A^{T} satisfies the identity $X^{\mathrm{T}}X = I$, so A^{T} is orthogonal.]

11. If A is orthogonal, prove that A^{-1} is orthogonal. [*Hint:* If A is orthogonal, then $A^{-1} = A^{\mathrm{T}}$.]

12. Let A, B be orthogonal $n \times n$ matrices. Show that the product AB is also an orthogonal $n \times n$ matrix. [*Hint:*

$$(AB)^{\mathrm{T}}(AB) = (B^{\mathrm{T}}A^{\mathrm{T}})(AB) = B^{\mathrm{T}}(A^{\mathrm{T}}A)B = B^{\mathrm{T}}B = I.]$$

13. Let $A^{n \times n}$ be orthogonal, and let $\mathbf{v}' = A\mathbf{v}$, where $\mathbf{v}$ is an $n \times 1$ column vector. Prove that $\mathbf{v} = A^{\mathrm{T}}\,\mathbf{v}'$.

*14. Let A be a real orthogonal $n \times 1$ matrix, and let $\{\mathbf{u}_1, \ldots, \mathbf{u}_n\}$ be an orthonormal set of $n \times 1$ column vectors in Euclidean n-space. Prove that $\{A\mathbf{u}_1, \ldots, A\mathbf{u}_n\}$ is also an orthonormal set of vectors.

14 SYMMETRIC MATRICES AND PRINCIPAL AXES

A *symmetric* matrix is a square matrix S for which $S^T = S$. Symmetric matrices arise naturally in the study of conic sections and quadric surfaces, as we shall soon see. Consider the problem of graphing the equation

(14.1) $$ax^2 + 2bxy + cy^2 + dx + ey + f = 0$$

in the XY-plane. Here, the coefficients $a, b, \ldots, f$ are real constants, with at least one of a, b, c not zero. The *graph* of the equation consists of all points (x,y) in the XY-plane whose coordinates satisfy the equation. The reader is probably familiar with the fact that the graph will be a conic section: parabola, ellipse, hyperbola, or some degenerate case such as a pair of straight lines, and so on. One way of sketching the graph is to introduce a new coordinate system, in which the equation of the conic section has no xy-term. Once this is done, we can use the simple technique of "completing the square" to reduce the equation to a standard form whose graph is easy to find.

In this problem, the main difficulty lies in dealing with the quadratic terms $ax^2 + 2bxy + cy^2$. With this expression we associate the symmetric matrix

$$S = \begin{bmatrix} a & b \\ b & c \end{bmatrix}.$$

Set $\mathbf{v} = \begin{bmatrix} x \\ y \end{bmatrix}$, so $\mathbf{v}^T = [x \quad y]$, and

(14.2) $$\mathbf{v}^T\, S\, \mathbf{v} = [x \quad y] S \begin{bmatrix} x \\ y \end{bmatrix} = ax^2 + 2bxy + cy^2.$$

As we shall see, we will be able to use the matrix S to graph equation (14.1).

Turning to three dimensions, consider the question of graphing the equation

(14.3) $$ax^2 + by^2 + cz^2 + 2dxy + 2exz + 2fyz + a'x + b'y + c'z + d' = 0,$$

where the coefficients are real numbers, and at least one quadratic term has a nonzero coefficient. The *graph* consists of all points (x,y,z) in XYZ-space whose coordinates satisfy (14.3). This graph will be a surface, such as an ellipsoid, paraboloid, hyperboloid, or some degenerate form thereof. Such surfaces are called *quadric surfaces*.

Again we shall concentrate on the quadratic terms

(14.4) $$ax^2 + by^2 + cz^2 + 2dxy + 2exz + 2fyz.$$

This expression is called a *quadratic form* in the variables x, y, z. With it we may associate the symmetric matrix

(14.5) $$S = \begin{bmatrix} a & d & e \\ d & b & f \\ e & f & c \end{bmatrix}.$$

If we put $\mathbf{v} = \begin{bmatrix} x \\ y \\ z \end{bmatrix}$, then the quadratic form (14.4) equals $\mathbf{v}^T S \mathbf{v}$. (We leave it to the reader to check this fact.)

Let us review a few ideas from Sections 11 and 12. Let S be a real symmetric $n \times n$ matrix. The *characteristic equation* of S is the equation $\det(\lambda I - S) = 0$. When we solve this equation for the unknown λ, we obtain n *characteristic roots* of S, say $\lambda_1, \ldots, \lambda_n$ (not necessarily distinct). Theorem 11.7 tells us that each λ_k is a real number. If $\mathbf{v}$ is a nonzero $n \times 1$ column vector for which

$$S\mathbf{v} = \lambda_k \mathbf{v},$$

we call $\mathbf{v}$ a *characteristic vector* of S belonging to the characteristic root λ_k. To determine such characteristic vectors, we first consider the *null space* of the matrix $\lambda_k I - S$. This null space consists of all vectors $\mathbf{v}$ such that

$$(\lambda_k I - S)\mathbf{v} = \mathbf{0}.$$

If the null space of $\lambda_k I - S$ is an r-dimensional vector space, we can find a basis $\{\mathbf{v}_1, \ldots, \mathbf{v}_r\}$ for this null space. Then $\{\mathbf{v}_1, \ldots, \mathbf{v}_r\}$ is a linearly independent set of vectors such that

(14.6) $$S\mathbf{v}_1 = \lambda_k \mathbf{v}_1, \ \ldots, \ S\mathbf{v}_r = \lambda_k \mathbf{v}_r.$$

Furthermore, because λ_k is real, it turns out that we can pick each of the vectors $\mathbf{v}_1, \ldots, \mathbf{v}_r$ to have real entries. Indeed, the following key result holds true:

(14.7) Theorem[1]

Let S be a real symmetric $n \times n$ matrix, and let λ_k be a characteristic root of S. Then

(i) *λ_k is a real number.*
(ii) *The null space of the matrix $\lambda_k I - S$ is a vector space whose dimension r equals the multiplicity of λ_k as a characteristic root of S.*
(iii) *There exists a basis $\{\mathbf{v}_1, \ldots, \mathbf{v}_r\}$ of the null space of $\lambda_k I - S$ such that each $\mathbf{v}_i$ is a real $n \times 1$ column vector. These vectors are characteristic vectors of S belonging to the characteristic root λ_k, that is, they satisfy equations* (14.6).

Continuing with our discussion, suppose that we have found vectors $\mathbf{v}_1, \ldots, \mathbf{v}_r$, as in (iii) above, which are a basis for the null space of $\lambda_k I - S$. We may then use the Gram-Schmidt method described in Section 12 to obtain an orthonormal basis $\{\mathbf{u}_1, \ldots, \mathbf{u}_r\}$ for this null space. In this way we find a set of mutually orthogonal unit vectors $\mathbf{u}_1, \ldots, \mathbf{u}_r$, satisfying

$$S\mathbf{u}_1 = \lambda_k \mathbf{u}_1, \quad \ldots, \quad S\mathbf{u}_r = \lambda_k \mathbf{u}_r.$$

Now suppose that λ_l is another characteristic root of S distinct from λ_k, and let λ_l occur with multiplicity s. As above, we may then find an orthonormal set of real vectors $\{\mathbf{w}_1, \ldots, \mathbf{w}_s\}$ such that

$$S\mathbf{w}_1 = \lambda_l \mathbf{w}_1, \quad \ldots, \quad S\mathbf{w}_s = \lambda_l \mathbf{w}_s.$$

The next theorem tells us that each $\mathbf{w}_i$ is orthogonal to each $\mathbf{u}_j$. Indeed, we have

(14.8) Theorem

Let λ, λ' be distinct characteristic roots of the real symmetrix matrix S. Let $\mathbf{u}$, $\mathbf{w}$ be vectors such that

$$S\mathbf{u} = \lambda \mathbf{u}, \qquad S\mathbf{w} = \lambda' \mathbf{w}.$$

Then $\mathbf{u} \cdot \mathbf{w} = 0$, that is, $\mathbf{w}$ is orthogonal to $\mathbf{u}$.

PROOF: See Exercise 14.6.

If we now combine the orthonormal sets $\{\mathbf{u}_1, \ldots, \mathbf{u}_r\}$ and

[1] For the proof, consult the references listed in the Preface. Part (i) has already been established in Theorem 11.7. The main step in the proof of (14.7) is to verify (ii). Part (iii) is an immediate consequence of parts (i) and (ii).

$\{\mathbf{w}_1, \ldots, \mathbf{w}_s\}$, we get a bigger set of mutually orthogonal unit vectors

$$\{\mathbf{u}_1, \ldots, \mathbf{u}_r, \mathbf{w}_1, \ldots, \mathbf{w}_s\}.$$

We thus have a new orthonormal set of vectors, each of which is a characteristic vector of S. In this set, the number of characteristic vectors belonging to a given characteristic root of S is exactly equal to the multiplicity of that characteristic root. Hence if we continue this procedure with all of the distinct characteristic roots of S, we will eventually obtain an orthonormal set of n vectors, each of which is a characteristic vector of S. We state this basic result explicitly as a theorem:

(14.9) Theorem

Let $S^{n\times n}$ be a real symmetric matrix, with characteristic roots $\lambda_1, \ldots, \lambda_n$. Then each λ_k is a real number, and there exists an orthonormal set $\{\mathbf{u}_1, \ldots, \mathbf{u}_n\}$ of real $n \times 1$ column vectors, such that

(14.10) $$S\mathbf{u}_1 = \lambda_1\mathbf{u}_1, \ldots, S\mathbf{u}_n = \lambda_n\mathbf{u}_n.$$

As a matter of fact, the preceding discussion shows us how to find such an orthonormal set $\{\mathbf{u}_1, \ldots, \mathbf{u}_n\}$, once the matrix S is given:

Step 1: Find the characteristic equation of S. This is the equation $\det(\lambda I - S) = 0$.

Step 2: Find the characteristic roots of S by solving the characteristic equation for λ.

Step 3: For each of the distinct characteristic roots λ_k of S, solve the equation

$$(\lambda_k I - S)\mathbf{v} = \mathbf{0}$$

for the vector $\mathbf{v}$. Use this solution to obtain a basis $\{\mathbf{v}_1, \ldots, \mathbf{v}_r\}$ for the null space of $\lambda_k I - S$, in which each $\mathbf{v}_i$ is a real vector. The number r should equal the multiplicity of λ_k as a characteristic root of S.

Step 4: Apply the Gram-Schmidt method (see Section 12) to find an orthonormal basis $\{\mathbf{u}_1, \ldots, \mathbf{u}_r\}$ for the null space of $\lambda_k I - S$. The $\mathbf{u}$'s are obtained successively by means of the formulas

$$\mathbf{u}_1 = \frac{1}{|\mathbf{v}_1|}\mathbf{v}_1,$$

$$\mathbf{v}_2' = \mathbf{v}_2 - (\mathbf{u}_1 \cdot \mathbf{v}_2)\mathbf{u}_1, \qquad \mathbf{u}_2 = \frac{1}{|\mathbf{v}_2'|}\mathbf{v}_2',$$

$$\mathbf{v}_3' = \mathbf{v}_3 - (\mathbf{u}_1 \cdot \mathbf{v}_3)\mathbf{u}_1 - (\mathbf{u}_2 \cdot \mathbf{v}_3)\mathbf{u}_2, \qquad \mathbf{u}_3 = \frac{1}{|\mathbf{v}_3'|}\mathbf{v}_3',$$

and so on.

Step 5: Perform Step 4 for each of the distinct characteristic roots λ_k of S, obtaining for each λ_k an orthonormal set of vectors belonging to that characteristic root λ_k. The desired orthonormal set $\{\mathbf{u}_1, \ldots, \mathbf{u}_n\}$ satisfying (14.10) is gotten by listing in succession the orthonormal sets associated with λ_k, where λ_k ranges over the distinct characteristic roots of S.

(14.11) EXAMPLE

Let $S = \begin{bmatrix} 3 & 1 \\ 1 & 3 \end{bmatrix}$,

$$\det(\lambda I - S) = \begin{vmatrix} \lambda - 3 & -1 \\ -1 & \lambda - 3 \end{vmatrix} = \lambda^2 - 6\lambda + 8.$$

The characteristic equation of S is $\lambda^2 - 6\lambda + 8 = 0$; solving for λ, we obtain two distinct roots $\lambda = 2$, $\lambda = 4$, each of multiplicity one.

(i) For the characteristic root $\lambda = 2$, we find the null space of $2I - S$ by solving the equation $(2I - S)\mathbf{v} = \mathbf{0}$ for $\mathbf{v}$. We have

$$(2I - S)\mathbf{v} = \begin{bmatrix} -1 & -1 \\ -1 & -1 \end{bmatrix} \begin{bmatrix} x \\ y \end{bmatrix} = \begin{bmatrix} -x - y \\ -x - y \end{bmatrix} = \begin{bmatrix} 0 \\ 0 \end{bmatrix},$$

getting

$$\begin{cases} -x - y = 0 \\ -x - y = 0. \end{cases}$$

These give $y = -x$, x arbitrary, so

$$\mathbf{v} = \begin{bmatrix} x \\ -x \end{bmatrix} = x \begin{bmatrix} 1 \\ -1 \end{bmatrix}.$$

The null space of $2I - S$ is 1-dimensional,[2] with basis vector $\mathbf{v}_1 = \begin{bmatrix} 1 \\ -1 \end{bmatrix}$.

The Gram-Schmidt procedure yields the unit vector

$$\mathbf{u}_1 = \frac{1}{|\mathbf{v}_1|}\mathbf{v}_1 = \frac{1}{\sqrt{2}} \begin{bmatrix} 1 \\ -1 \end{bmatrix} = \begin{bmatrix} \frac{1}{\sqrt{2}} \\ \frac{-1}{\sqrt{2}} \end{bmatrix},$$

and we have $S\mathbf{u}_1 = 2\mathbf{u}_1$.

(ii) Repeat the procedure for the characteristic root $\lambda = 4$:

[2] This is as expected, since $\lambda = 2$ is a characteristic root of multiplicity one.

$$(4I - S)\mathbf{v} = \begin{bmatrix} 1 & -1 \\ -1 & 1 \end{bmatrix}\begin{bmatrix} x \\ y \end{bmatrix} = \begin{bmatrix} x - y \\ -x + y \end{bmatrix} = \begin{bmatrix} 0 \\ 0 \end{bmatrix},$$

so

$$\begin{cases} x - y = 0 \\ -x + y = 0. \end{cases}$$

Then $x = y$, y arbitrary, and $\mathbf{v} = \begin{bmatrix} x \\ x \end{bmatrix} = x\begin{bmatrix} 1 \\ 1 \end{bmatrix}$. The null space of $4I - S$ is 1-dimensional (as expected), with basis vector $\mathbf{v}_2 = \begin{bmatrix} 1 \\ 1 \end{bmatrix}$. A unit basis vector is $\mathbf{u}_2 = \frac{1}{\sqrt{2}}\begin{bmatrix} 1 \\ 1 \end{bmatrix}$, and $S\mathbf{u}_2 = 4\mathbf{u}_2$. As a check on our calculations, we verify that $\mathbf{u}_1 \cdot \mathbf{u}_2 = 0$:

$$\mathbf{u}_1 \cdot \mathbf{u}_2 = \left(\frac{1}{\sqrt{2}}\right)\left(\frac{1}{\sqrt{2}}\right) + \left(\frac{-1}{\sqrt{2}}\right)\left(\frac{1}{\sqrt{2}}\right) = 0.$$

(This is as predicted by Theorem 14.8.)

(14.12) EXAMPLE

Given $S = \begin{bmatrix} 7 & -1 & -2 \\ -1 & 7 & 2 \\ -2 & 2 & 10 \end{bmatrix}$, we have

$$\det(\lambda I - S) = \begin{vmatrix} \lambda - 7 & 1 & 2 \\ 1 & \lambda - 7 & -2 \\ 2 & -2 & \lambda - 10 \end{vmatrix} = \lambda^3 - 24\lambda^2 + 180\lambda - 432$$

$$= (\lambda - 6)(\lambda - 6)(\lambda - 12).$$

(i) For the characteristic root $\lambda = 6$, find the null space of $6I - S$:

$$(6I - S)\mathbf{v} = \begin{bmatrix} -1 & 1 & 2 \\ 1 & -1 & -2 \\ 2 & -2 & -4 \end{bmatrix}\begin{bmatrix} x \\ y \\ z \end{bmatrix} = \begin{bmatrix} -x + y + 2z \\ x - y - 2z \\ 2x - 2y - 4z \end{bmatrix},$$

$$\begin{cases} -x + y + 2z = 0 \\ x - y - 2z = 0 \\ 2x - 2y - 4z = 0 \end{cases} \xrightarrow[r_3 + 2r_1]{r_2 + r_1} \begin{cases} -x + y + 2z = 0 \\ 0 = 0 \\ 0 = 0, \end{cases}$$

so $x = y + 2z$, where y and z are arbitrary. Then

$$\mathbf{v} = \begin{bmatrix} y + 2z \\ y \\ z \end{bmatrix} = y\begin{bmatrix} 1 \\ 1 \\ 0 \end{bmatrix} + z\begin{bmatrix} 2 \\ 0 \\ 1 \end{bmatrix},$$

so the null space of $6I - S$ has basis $\{\mathbf{v}_1, \mathbf{v}_2\}$ given by

$$\mathbf{v}_1 = \begin{bmatrix} 1 \\ 1 \\ 0 \end{bmatrix}, \qquad \mathbf{v}_2 = \begin{bmatrix} 2 \\ 0 \\ 1 \end{bmatrix}.$$

Use the Gram-Schmidt method:

$$\mathbf{u}_1 = \frac{1}{|\mathbf{v}_1|}\mathbf{v}_1 = \frac{1}{\sqrt{2}}\begin{bmatrix} 1 \\ 1 \\ 0 \end{bmatrix}, \qquad \mathbf{u}_1 \cdot \mathbf{v}_2 = \frac{2}{\sqrt{2}} = \sqrt{2},$$

$$\mathbf{v}_2' = \mathbf{v}_2 - (\mathbf{u}_1 \cdot \mathbf{v}_2)\mathbf{u}_1 = \begin{bmatrix} 2 \\ 0 \\ 1 \end{bmatrix} - \sqrt{2} \cdot \frac{1}{\sqrt{2}}\begin{bmatrix} 1 \\ 1 \\ 0 \end{bmatrix} = \begin{bmatrix} 1 \\ -1 \\ 1 \end{bmatrix},$$

$$\mathbf{u}_2 = \frac{1}{|\mathbf{v}_2'|}\mathbf{v}_2 = \frac{1}{\sqrt{3}}\begin{bmatrix} 1 \\ -1 \\ 1 \end{bmatrix}.$$

Then $\{\mathbf{u}_1, \mathbf{u}_2\}$ is an orthonormal set, and $S\mathbf{u}_1 = 6\mathbf{u}_1$, $S\mathbf{u}_2 = 6\mathbf{u}_2$. (Note that 6 is a characteristic root of multiplicity 2.)
(ii) For $\lambda = 12$, we obtain

$$(12I - S)\mathbf{v} = \begin{bmatrix} 5 & 1 & 2 \\ 1 & 5 & -2 \\ 2 & -2 & 2 \end{bmatrix}\begin{bmatrix} x \\ y \\ z \end{bmatrix} = \begin{bmatrix} 0 \\ 0 \\ 0 \end{bmatrix},$$

so

$$\begin{cases} 5x + y + 2z = 0 \\ x + 5y - 2z = 0 \\ 2x - 2y + 2z = 0 \end{cases} \xrightarrow[r_3 - 2r_2]{r_1 - 5r_2} \begin{cases} -24y + 12z = 0 \\ x + 5y - 2z = 0 \\ -12y + 6z = 0 \end{cases}$$

$$\xrightarrow[\text{rearrange}]{r_1 - 2r_3} \begin{cases} x + 5y - 2z = 0 \\ -12y + 6z = 0 \\ 0 = 0 \end{cases}$$

$$\longrightarrow \begin{cases} z = 2y, \\ x = -y, \ y \text{ arbitrary.} \end{cases}$$

Then

$$\mathbf{v} = \begin{bmatrix} -y \\ y \\ 2y \end{bmatrix} = y\begin{bmatrix} -1 \\ 1 \\ 2 \end{bmatrix}, \qquad \mathbf{u}_3 = \frac{1}{\sqrt{6}}\begin{bmatrix} -1 \\ 1 \\ 2 \end{bmatrix}.$$

Therefore $\mathbf{u}_3$ is a unit vector such that $S\mathbf{u}_3 = 12\mathbf{u}_3$. We leave it to the reader to check that $\mathbf{u}_1 \cdot \mathbf{u}_3 = \mathbf{u}_2 \cdot \mathbf{u}_3 = 0$, as expected from Theorem 14.8.

As we shall see below, the procedure illustrated above can be used to eliminate cross-product terms in equations of conic sections and quadric surfaces. Consider for example the equation

$$ax^2 + 2bxy + cy^2 = k,$$

where a, b, c, k are real constants, and at least one of a, b, c is not zero. Its graph (see Figure 14.1) is some conic section C in the XY-plane (or possibly some degenerate conic, such as a point or a pair of lines). We wish to introduce new axes OX', OY' in the XY-plane such that in the $X'Y'$-coordinate system, the equation of C is of the form

$$a'(x')^2 + c'(y')^2 = k.$$

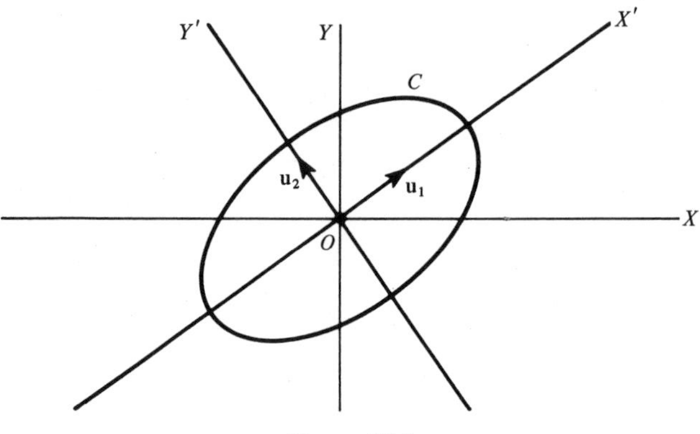

Figure 14.1

The fundamental result in this direction is as follows:

(14.13) Principal Axes Theorem

(i) (*2-dimensional case.*) *Let the conic section C be the graph of the equation $ax^2 + 2bxy + cy^2 = k$ in the XY-plane, and also let* $S = \begin{bmatrix} a & b \\ b & c \end{bmatrix}$ *be the real symmetric matrix that is associated with the quadratic form $ax^2 + 2bxy + cy^2$. Suppose that $\{\mathbf{u}_1, \mathbf{u}_2\}$ is an orthonormal set of vectors in the XY-plane, chosen so that*[3]

$$S\mathbf{u}_1 = \lambda_1\mathbf{u}_1, \qquad S\mathbf{u}_2 = \lambda_2\mathbf{u}_2,$$

where λ_1, λ_2 are the characteristic roots of S. Introduce new axes OX', OY' in the XY-plane, in the directions $\mathbf{u}_1$, $\mathbf{u}_2$, respectively. Then the equation of C in the $X'Y'$-coordinate system is

$$\lambda_1(x')^2 + \lambda_2(y')^2 = k.$$

[3] Such a choice is always possible, by Theorem 14.9.

(ii) (*3-dimensional case.*) *Let Q be the quadric surface defined by*

$$ax^2 + by^2 + cz^2 + 2dxy + 2exz + 2fyz = k,$$

and let S be the associated real symmetric matrix given by (14.5). *Let $\{\mathbf{u}_1, \mathbf{u}_2, \mathbf{u}_3\}$ be an orthonormal set of vectors in XYZ-space, chosen so that*[4]

$$S\mathbf{u}_1 = \lambda_1\mathbf{u}_1, \qquad S\mathbf{u}_2 = \lambda_2\mathbf{u}_2, \qquad S\mathbf{u}_3 = \lambda_3\mathbf{u}_3,$$

where $\lambda_1, \lambda_2, \lambda_3$ are the characteristic roots of S. Introduce new axes OX', OY', OZ' in the directions $\mathbf{u}_1, \mathbf{u}_2, \mathbf{u}_3$, respectively. Then the equation of Q in the $X'Y'Z'$-coordinate system is

$$\lambda_1(x')^2 + \lambda_2(y')^2 + \lambda_3(z')^2 = k.$$

*[*PROOF:* For simplicity, consider only case (i). Let (x,y) denote the XY-coordinates of an arbitrary point P in the XY-plane, and let (x',y') be the $X'Y'$-coordinates of this same point P. Let $\mathbf{i}, \mathbf{j}$ be unit vectors along OX, OY, respectively, and represent the vector $\mathbf{v} = \overrightarrow{OP}$ as a column vector $\begin{bmatrix} x \\ y \end{bmatrix}$. (See Figure 14.2.)

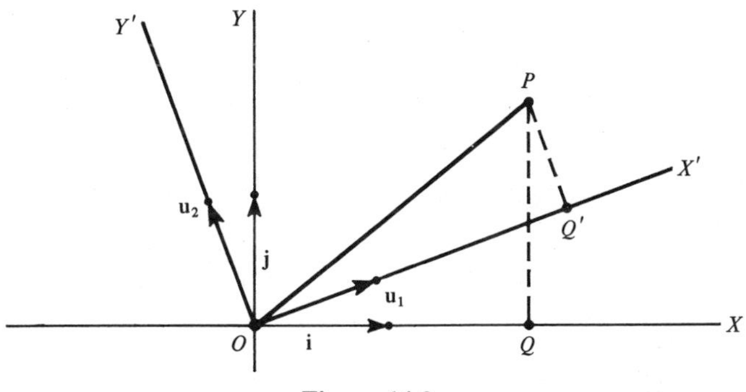

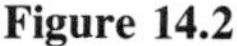
Figure 14.2

Clearly we have $\overrightarrow{OQ} = x\mathbf{i}$, $\overrightarrow{QP} = y\mathbf{j}$, and so

$$\mathbf{v} = \overrightarrow{OP} = \overrightarrow{OQ} + \overrightarrow{QP} = x\mathbf{i} + y\mathbf{j}.$$

On the other hand, $\overrightarrow{OQ'} = x'\mathbf{u}_1$, $\overrightarrow{Q'P} = y'\mathbf{u}_2$, which gives

(14.14) $$\mathbf{v} = \overrightarrow{OP} = \overrightarrow{OQ'} + \overrightarrow{Q'P} = x'\mathbf{u}_1 + y'\mathbf{u}_2.$$

We have seen in (14.2) that the equation of the conic C may be

[4] Such a choice is always possible, by Theorem 14.9.

written in the form

$$\textbf{(14.15)} \qquad \mathbf{v}^{\mathrm{T}} S \, \mathbf{v} = k.$$

From (14.14) we may calculate $\mathbf{v}^{\mathrm{T}}$, namely,

$$\mathbf{v}^{\mathrm{T}} = (x'\mathbf{u}_1 + y'\mathbf{u}_2)^{\mathrm{T}} = x'\mathbf{u}_1{}^{\mathrm{T}} + y'\mathbf{u}_2{}^{\mathrm{T}}.$$

Therefore (14.15) may be rewritten as

$$\textbf{(14.16)} \qquad (x'\mathbf{u}_1{}^{\mathrm{T}} + y'\mathbf{u}_2{}^{\mathrm{T}}) S (x'\mathbf{u}_1 + y'\mathbf{u}_2) = k.$$

This is now the equation of the conic C relative to the $X'Y'$-coordinate system. We have not changed the curve C in any way; rather, we have a new description of C as the set of all points P in the $X'Y'$-plane whose $X'Y'$-coordinates (x',y') satisfy (14.16).

Now let us use our hypotheses that $\{\mathbf{u}_1, \mathbf{u}_2\}$ form an orthonormal set of vectors satisfying $S\mathbf{u}_1 = \lambda_1\mathbf{u}_1$, $S\mathbf{u}_2 = \lambda_2\mathbf{u}_2$. We shall simplify equation (14.16) by using these conditions on $\mathbf{u}_1$ and $\mathbf{u}_2$. Indeed, we may rewrite (14.16) as

$$(x'\mathbf{u}_1{}^{\mathrm{T}} + y'\mathbf{u}_2{}^{\mathrm{T}})(x'S\mathbf{u}_1 + y'S\mathbf{u}_2) = k,$$

that is,

$$(x'\mathbf{u}_1{}^{\mathrm{T}} + y'\mathbf{u}_2{}^{\mathrm{T}})(x'\lambda_1\mathbf{u}_1 + y'\lambda_2\mathbf{u}_2) = k.$$

Multiplying out, we find that the equation of C in the $X'Y'$-coordinate system is

$$\textbf{(14.17)} \quad \lambda_1 x'^2 \, \mathbf{u}_1{}^{\mathrm{T}} \mathbf{u}_1 + x'y'(\lambda_1\mathbf{u}_2{}^{\mathrm{T}} \mathbf{u}_1 + \lambda_2\mathbf{u}_1{}^{\mathrm{T}} \mathbf{u}_2) + \lambda_2 y'^2 \, \mathbf{u}_2{}^{\mathrm{T}} \mathbf{u}_2 = k.$$

We are now ready to use the identity (13.6), which tells us that

$$\mathbf{u}_1{}^{\mathrm{T}} \mathbf{u}_1 = \mathbf{u}_1 \cdot \mathbf{u}_1, \qquad \mathbf{u}_2{}^{\mathrm{T}} \mathbf{u}_1 = \mathbf{u}_2 \cdot \mathbf{u}_1,$$

and so on. Since $\{\mathbf{u}_1, \mathbf{u}_2\}$ is an orthonormal set, we know that

$$\mathbf{u}_1 \cdot \mathbf{u}_1 = 1, \qquad \mathbf{u}_2 \cdot \mathbf{u}_1 = 0, \qquad \mathbf{u}_2 \cdot \mathbf{u}_2 = 1.$$

Therefore we have

$$\mathbf{u}_1{}^{\mathrm{T}} \mathbf{u}_1 = 1, \qquad \mathbf{u}_2{}^{\mathrm{T}} \mathbf{u}_1 = 0, \qquad \mathbf{u}_1{}^{\mathrm{T}} \mathbf{u}_2 = 0, \qquad \mathbf{u}_2{}^{\mathrm{T}} \mathbf{u}_2 = 1.$$

Equation (14.17) thus reduces to the simple form

$$\lambda_1 x'^2 + \lambda_2 y'^2 = k.$$

This is the equation of C in the $X'Y'$-coordinate system. Thus the proof of the Principal Axes Theorem in the 2-dimensional case is completed.]*

Remarks:

1. The axes of symmetry of the conic C are OX', OY'. These are called the *principal axes* of C.

2. What is the relation between the XY-coordinates (x,y) and the $X'Y'$-coordinates (x',y') of an arbitrary point P in the XY-plane? Write $\mathbf{u}_1 = \overrightarrow{OP}_1$, $\mathbf{u}_2 = \overrightarrow{OP}_2$, where (x_1,y_1) are the XY-coordinates of P_1, and (x_2,y_2) are those of P_2. Then $\mathbf{u}_1 = \begin{bmatrix} x_1 \\ y_1 \end{bmatrix}$, $\mathbf{u}_2 = \begin{bmatrix} x_2 \\ y_2 \end{bmatrix}$, and (14.14) gives

$$\begin{bmatrix} x \\ y \end{bmatrix} = x' \begin{bmatrix} x_1 \\ y_1 \end{bmatrix} + y' \begin{bmatrix} x_2 \\ y_2 \end{bmatrix} = \begin{bmatrix} x_1 & x_2 \\ y_1 & y_2 \end{bmatrix} \begin{bmatrix} x' \\ y' \end{bmatrix}.$$

The matrix $U = \begin{bmatrix} x_1 & x_2 \\ y_1 & y_2 \end{bmatrix}$ has column vectors $\mathbf{u}_1$, $\mathbf{u}_2$; these form an orthonormal set, so by (13.3) U is an orthogonal matrix (and thus $U^{-1} = U^{\mathrm{T}}$). Therefore we have

(14.18)
$$\begin{bmatrix} x \\ y \end{bmatrix} = U \begin{bmatrix} x' \\ y' \end{bmatrix}, \qquad \begin{bmatrix} x' \\ y' \end{bmatrix} = U^{-1} \begin{bmatrix} x \\ y \end{bmatrix} = U^{\mathrm{T}} \begin{bmatrix} x \\ y \end{bmatrix}.$$

In the 3-dimensional case, let U be the 3×3 orthogonal matrix with columns $\mathbf{u}_1$, $\mathbf{u}_2$, $\mathbf{u}_3$. Then, analogously, the relation between XYZ- and $X'Y'Z'$-coordinates is given by

(14.19)
$$\begin{bmatrix} x \\ y \\ z \end{bmatrix} = U \begin{bmatrix} x' \\ y' \\ z' \end{bmatrix}, \qquad \begin{bmatrix} x' \\ y' \\ z' \end{bmatrix} = U^{\mathrm{T}} \begin{bmatrix} x \\ y \\ z \end{bmatrix}.$$

3. The preceding remark suggests another point of view: the orthogonal transformation of the XY-plane, defined by letting $\overrightarrow{OP}$ map onto $U^{-1} \cdot \overrightarrow{OP}$, carries the conic C onto a congruent conic C' with principal axes OX, OY. (We shall *not* adopt this point of view here, however, and shall instead keep fixed the curve C, and introduce a second coordinate system as above.)

(14.20) EXAMPLE

Consider the conic C whose equation is $3x^2 + 2xy + 3y^2 = 4$. The symmetric matrix S associated with the quadratic form $3x^2 + 2xy + 3y^2$ is given by $S = \begin{bmatrix} 3 & 1 \\ 1 & 3 \end{bmatrix}$. We have already seen in (14.11) that

$$S\mathbf{u}_1 = 2\mathbf{u}_1, \qquad S\mathbf{u}_2 = 4\mathbf{u}_2,$$

where

$$\mathbf{u}_1 = \begin{bmatrix} \frac{1}{\sqrt{2}} \\ \frac{-1}{\sqrt{2}} \end{bmatrix}, \qquad \mathbf{u}_2 = \begin{bmatrix} \frac{1}{\sqrt{2}} \\ \frac{1}{\sqrt{2}} \end{bmatrix}.$$

Let OX', OY' be new axes in the directions $\mathbf{u}_1$, $\mathbf{u}_2$, respectively. By the

Principal Axes Theorem, the equation of C in the $X'Y'$-coordinate system is

$$2x'^2 + 4y'^2 = 4.$$

(In this case, $\lambda_1 = 2$, $\lambda_2 = 4$, $k = 4$.) The above equation can be written as

$$\frac{x'^2}{2} + \frac{y'^2}{1} = 1.$$

Its graph, Figure 14.3, is an ellipse with semimajor axis OA of length $\sqrt{2}$, semiminor axis OB of length 1. The axes of symmetry of C are OX' and OY'.

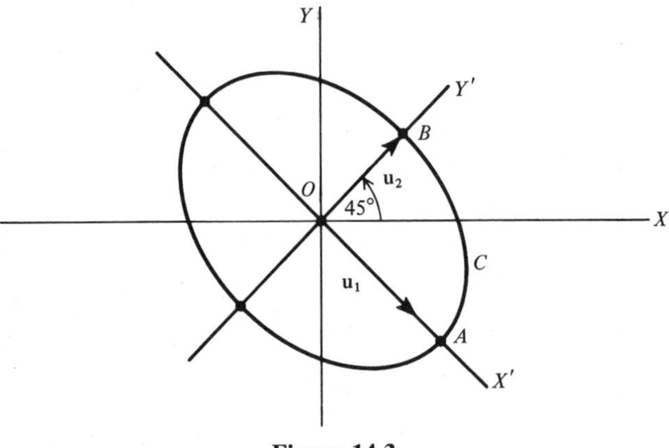

Figure 14.3

Let us use (14.18) to obtain the relations between the XY- and $X'Y'$-coordinate systems. The matrix U whose columns are $\mathbf{u}_1$, $\mathbf{u}_2$, is given in this case by

$$U = \frac{1}{\sqrt{2}}\begin{bmatrix} 1 & 1 \\ -1 & 1 \end{bmatrix}.$$

Therefore

(14.21) $$\begin{bmatrix} x \\ y \end{bmatrix} = \frac{1}{\sqrt{2}}\begin{bmatrix} 1 & 1 \\ -1 & 1 \end{bmatrix}\begin{bmatrix} x' \\ y' \end{bmatrix}, \quad \text{that is,} \quad \begin{cases} x = \dfrac{1}{\sqrt{2}}(x' + y') \\ y = \dfrac{1}{\sqrt{2}}(-x' + y'). \end{cases}$$

(14.22) EXAMPLE

Let us consider the conic C whose equation is

$$7x^2 - 48xy - 7y^2 = 100.$$

In this case

$$S = \begin{bmatrix} 7 & -24 \\ -24 & -7 \end{bmatrix}.$$

Following the procedure of example (14.11), we obtain

$$\det(\lambda I - S) = \begin{vmatrix} \lambda - 7 & 24 \\ 24 & \lambda + 7 \end{vmatrix} = \lambda^2 - 625,$$

so S has characteristic roots $\lambda_1 = 25$, $\lambda_2 = -25$.

Next, the condition

$$(\lambda_1 I - S)\mathbf{v} = \begin{bmatrix} 18 & 24 \\ 24 & 32 \end{bmatrix}\begin{bmatrix} x \\ y \end{bmatrix} = \begin{bmatrix} 18x + 24y \\ 24x + 32y \end{bmatrix} = \begin{bmatrix} 0 \\ 0 \end{bmatrix}$$

gives $18x + 24y = 0$, $24x + 32y = 0$. Thus $x = \left(\frac{-4}{3}\right) y$, y arbitrary. Taking $y = -3$ for convenience, we see that the null space of $\lambda_1 I - S$ has basis $\begin{bmatrix} 4 \\ -3 \end{bmatrix}$; therefore we obtain

$$\mathbf{u}_1 = \frac{1}{5}\begin{bmatrix} 4 \\ -3 \end{bmatrix}, \qquad S\mathbf{u}_1 = 25\mathbf{u}_1.$$

Likewise, solving

$$(\lambda_2 I - S)\mathbf{v} = \begin{bmatrix} -32 & 24 \\ 24 & -18 \end{bmatrix}\begin{bmatrix} x \\ y \end{bmatrix} = \begin{bmatrix} 0 \\ 0 \end{bmatrix},$$

we find a basis vector $\begin{bmatrix} 3 \\ 4 \end{bmatrix}$ for the null space of $\lambda_2 I - S$. Hence we may put

$$\mathbf{u}_2 = \frac{1}{5}\begin{bmatrix} 3 \\ 4 \end{bmatrix}, \qquad S\mathbf{u}_2 = -25\mathbf{u}_2.$$

The equation of C in the $X'Y'$-coordinate system is

$$25x'^2 - 25y'^2 = 100,$$

that is,

$$\frac{x'^2}{4} - \frac{y'^2}{4} = 1.$$

Therefore C is a hyperbola with principal axes OX', OY'.

From (14.18) we obtain

$$\begin{bmatrix} x \\ y \end{bmatrix} = \frac{1}{5}\begin{bmatrix} 4 & 3 \\ -3 & 4 \end{bmatrix}\begin{bmatrix} x' \\ y' \end{bmatrix}, \qquad \begin{bmatrix} x' \\ y' \end{bmatrix} = \frac{1}{5}\begin{bmatrix} 4 & -3 \\ 3 & 4 \end{bmatrix}\begin{bmatrix} x \\ y \end{bmatrix}.$$

The geometric picture, Figure 14.4, is as follows (note that $OA = OA' = 2$):

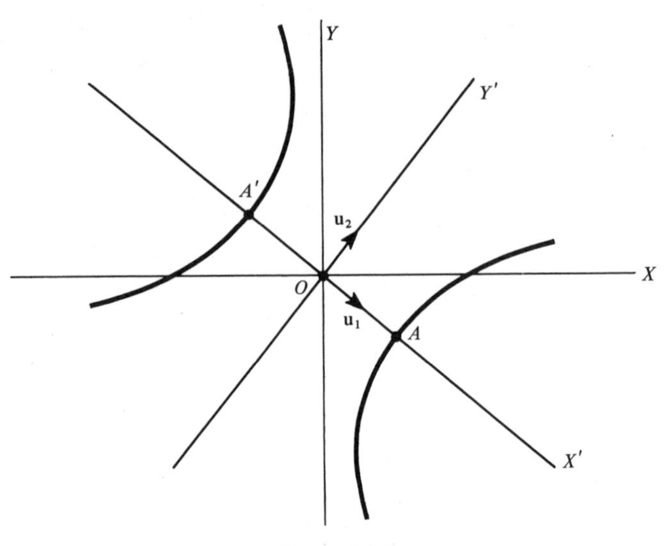

Figure 14.4

The asymptotes of the hyperbola are given by $y' = \pm x'$, that is, by the equations

$$\tfrac{1}{5}(3x + 4y) = \pm\tfrac{1}{5}(4x - 3y).$$

Thus the asymptotes are the lines $x = 7y$ and $y = -7x$. (Indeed, the original equation is expressible as

$$(7x + y)(x - 7y) = 100.$$

The asymptotes are obtained by separately setting each linear factor equal to zero.)

(14.23) EXAMPLE

Consider the quadric surface Q whose equation is

$$7x^2 + 7y^2 + 10z^2 - 2xy - 4xz + 4yz = 24.$$

The corresponding symmetric matrix S was already studied in (14.12). Its characteristic roots are $\lambda_1 = 6$, $\lambda_2 = 6$, $\lambda_3 = 12$, and we found that

$$S\mathbf{u}_1 = 6\mathbf{u}_1, \qquad S\mathbf{u}_2 = 6\mathbf{u}_2, \qquad S\mathbf{u}_3 = 12\mathbf{u}_3,$$

where $\{\mathbf{u}_1, \mathbf{u}_2, \mathbf{u}_3\}$ is the orthonormal set:

$$\mathbf{u}_1 = \frac{1}{\sqrt{2}}\begin{bmatrix} 1 \\ 1 \\ 0 \end{bmatrix}, \qquad \mathbf{u}_2 = \frac{1}{\sqrt{3}}\begin{bmatrix} 1 \\ -1 \\ 1 \end{bmatrix}, \qquad \mathbf{u}_3 = \frac{1}{\sqrt{6}}\begin{bmatrix} -1 \\ 1 \\ 2 \end{bmatrix}.$$

Let OX', OY', OZ' be new axes in space in the directions $\mathbf{u}_1$, $\mathbf{u}_2$, $\mathbf{u}_3$, respectively. These are principal axes of the quadric surface Q; in the new $X'Y'Z'$-coordinate system, the equation of Q is

$$6x'^2 + 6y'^2 + 12z'^2 = 24,$$

that is,

$$\frac{x'^2}{4} + \frac{y'^2}{4} + \frac{z'^2}{2} = 1.$$

Thus Q is an ellipsoid of revolution about the axis OZ', and its semi-axes are of lengths 2, 2, $\sqrt{2}$, respectively. Note that we could have chosen another orthonormal set $\{\mathbf{u}_1', \mathbf{u}_2'\}$ such that $S\mathbf{u}_1' = 6\mathbf{u}_1'$, $S\mathbf{u}_2' = 6\mathbf{u}_2'$; indeed, this corresponds to the fact that the axes OX', OY' are not uniquely determined relative to the surface Q. We could have used in place of OX', OY' any pair of mutually perpendicular axes OX'', OY'' in the plane $OX'Y'$.

Figure 14.5 is a sketch of the ellipsoid Q, and the new system of coordinate axes OX', OY', OZ'. We have distorted distances somewhat, in order to make the picture look nicer. Note that $OA = 2$, $OB = 2$, $OC = \sqrt{2}$.

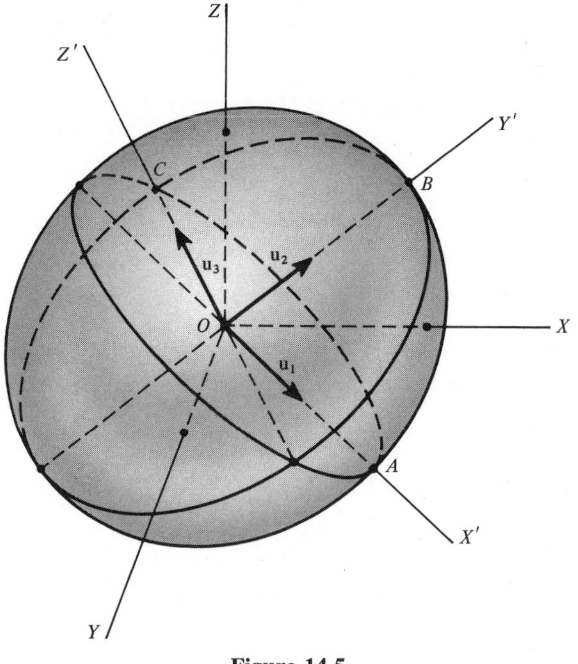

Figure 14.5

We conclude with an example showing how to deal with equation (14.1) when linear and quadratic terms both occur.

(14.24) EXAMPLE

Consider the conic C whose equation is

$$3x^2 + 2xy + 3y^2 + 4\sqrt{2}x - 8\sqrt{2}y + 11 = 0.$$

Introduce new axes as in (14.20). From (14.21) we have

$$x = \frac{1}{\sqrt{2}}(x' + y'), \qquad y = \frac{1}{\sqrt{2}}(-x' + y').$$

We have already seen in Example 14.20 that

$$3x^2 + 2xy + 3y^2 = 2x'^2 + 4y'^2,$$

so the equation of C relative to the $X'Y'$-coordinate system becomes

$$2x'^2 + 4y'^2 + 4\sqrt{2} \cdot \frac{1}{\sqrt{2}}(x' + y') - 8\sqrt{2} \cdot \frac{1}{\sqrt{2}}(-x' + y') + 11 = 0.$$

Simplifying, we obtain

$$2x'^2 + 12x' + 4y'^2 - 4y' + 11 = 0.$$

Write this as

(14.25) $$2\{x'^2 + 6x' + ?\} + 4\{y'^2 - y' + ?\} = -11.$$

Now observe that

$$x^2 + rx + (\tfrac{1}{2}r)^2 = (x + \tfrac{1}{2}r)^2.$$

Thus

$$x'^2 + 6x' + 9 = (x' + 3)^2, \qquad y'^2 - y' + \tfrac{1}{4} = (y' - \tfrac{1}{2})^2.$$

Replacing the question marks in (14.25) by 9 and $\frac{1}{4}$, respectively, that equation becomes

$$2\{x'^2 + 6x' + 9\} + 4\{y'^2 - y' + \tfrac{1}{4}\} = -11 + 2 \cdot 9 + 4 \cdot \tfrac{1}{4}.$$

We rewrite the equation as follows:

$$2(x' + 3)^2 + 4(y' - \tfrac{1}{2})^2 = 8.$$

Now set

$$x'' = x' + 3, \qquad y'' = y' - \tfrac{1}{2},$$

and introduce $X''Y''$-axes. The origin O'' has coordinates $(-3, \frac{1}{2})$ relative to the $X'Y'$-axes (see Figure 14.6):

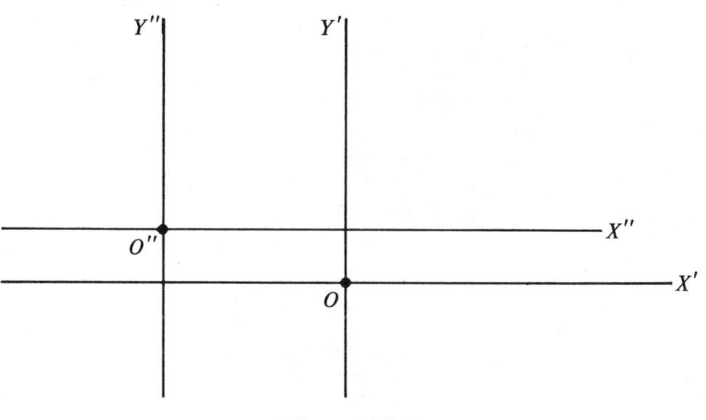

Figure 14.6

The equation of the conic C in the $X''Y''$-coordinate system is

$$2(x'')^2 + 4(y'')^2 = 8,$$

and $O''X''$, $O''Y''$ are the principal axes of C (see Figure 14.7).

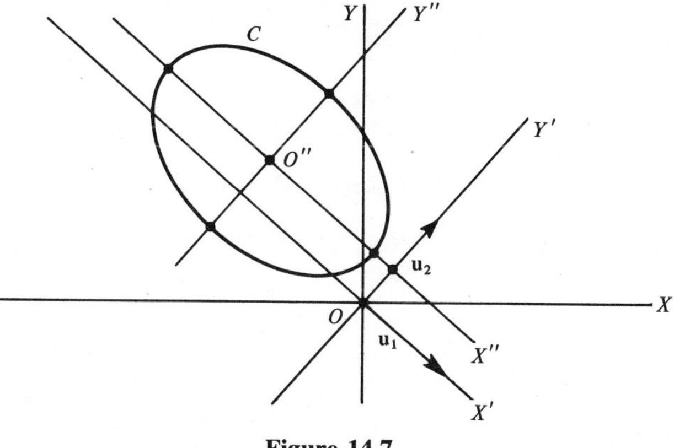

Figure 14.7

EXERCISES

1. For each of the following conics, find axes of symmetry and sketch the curve:
   (a) $x^2 + 4y^2 - 6x + 16y + 9 = 0$.
   (b) $x^2 - 9y^2 + 18y - 18 = 0$.
   (c) $4x^2 + 2y + 5 = 0$.
2. Find principal axes and sketch for various values of the constant k:
   (a) $x^2 + 2xy - 3y^2 = k$.
   (b) $x^2 + 4xy + y^2 = k$.

(c) $2x^2 - 4xy + 5y^2 = k$.
(d) $x^2 + 2xy + y^2 = 9$.
(e) $x^2 + 2xy + y^2 + \sqrt{2}(x - y) + 6 = 0$.
(f) $x^2 + 4xy + 4y^2 + 10x = 0$.

3. Find principal axes of the following quadric surfaces, and sketch for various values of the constant k:
(a) $x^2 + 2(y^2 + yz + z^2) = k$.
(b) $2(xy + xz + yz) = 8$.

4. Let

$$S = \begin{bmatrix} 25 & 16 & -8 \\ 16 & 25 & -8 \\ -8 & -8 & 13 \end{bmatrix}.$$

Check that S has characteristic roots 9, 9, and 45. Find an orthonormal set of 3×1 column vectors $\{\mathbf{u}_1, \mathbf{u}_2, \mathbf{u}_3\}$ such that

$$S\mathbf{u}_1 = 9\mathbf{u}_1, \qquad S\mathbf{u}_2 = 9\mathbf{u}_2, \qquad S\mathbf{u}_3 = 45\mathbf{u}_3.$$

*5. Let S be the matrix given in Example 14.12. Find a basis for the null space of $6I - S$, other than the basis obtained in that example. Find another pair of mutually orthogonal unit vectors $\mathbf{u}_1'$, $\mathbf{u}_2'$ such that $S\mathbf{u}_1' = 6\mathbf{u}_1'$, $S\mathbf{u}_2' = 6\mathbf{u}_2'$.

*6. Let λ, λ' be a pair of distinct characteristic roots of the real symmetric matrix S, and let $\mathbf{u}$, $\mathbf{v}$ be vectors such that

$$S\mathbf{u} = \lambda\mathbf{u}, \qquad S\mathbf{v} = \lambda'\mathbf{v}.$$

Prove that $\mathbf{u} \cdot \mathbf{v} = 0$. [*Hint:* By (13.6), we need only show that $\mathbf{u}^T\mathbf{v} = 0$. From $S\mathbf{u} = \lambda\mathbf{u}$ we get $\mathbf{u}^T S^T = \lambda\mathbf{u}^T$. But $S^T = S$, so $\mathbf{u}^T S = \lambda\mathbf{u}^T$, and therefore

$$(\mathbf{u}^T S)\mathbf{v} = \lambda\mathbf{u}^T\mathbf{v}.$$

But also

$$(\mathbf{u}^T S)\mathbf{v} = \mathbf{u}^T(S\mathbf{v}) = \mathbf{u}^T(\lambda'\mathbf{v}) = \lambda'\mathbf{u}^T\mathbf{v}.$$

Therefore $(\lambda - \lambda')\mathbf{u}^T\mathbf{v} = 0$, so $\mathbf{u}^T\mathbf{v} = 0$.]

*7. Let λ_1, λ_2 be the characteristic roots of the real symmetric 2×2 matrix S, and let $\{\mathbf{u}_1, \mathbf{u}_2\}$ be an orthonormal set of vectors such that

$$S\mathbf{u}_1 = \lambda_1\mathbf{u}_1, \qquad S\mathbf{u}_2 = \lambda_2\mathbf{u}_2.$$

Let $U = [\mathbf{u}_1 \;\; \mathbf{u}_2]$ be the matrix with column vectors $\mathbf{u}_1$, $\mathbf{u}_2$. Show that U is an orthogonal matrix, and that

$$U^T S U = \begin{bmatrix} \lambda_1 & 0 \\ 0 & \lambda_2 \end{bmatrix}.$$

Can you generalize this to the 3×3 case? [*Hint:* U is orthogonal by (13.3). Next,

$$SU = S[\mathbf{u}_1 \quad \mathbf{u}_2] = [S\mathbf{u}_1 \quad S\mathbf{u}_2] = [\lambda_1\mathbf{u}_1 \quad \lambda_2\mathbf{u}_2].$$

Therefore

$$U^{\mathrm{T}}SU = \begin{bmatrix} \mathbf{u}_1{}^{\mathrm{T}} \\ \mathbf{u}_2{}^{\mathrm{T}} \end{bmatrix} [\lambda_1\mathbf{u}_1 \quad \lambda_2\mathbf{u}_2] \quad \begin{bmatrix} \lambda_1\mathbf{u}_1{}^{\mathrm{T}}\,\mathbf{u}_1 & \lambda_2\mathbf{u}_1{}^{\mathrm{T}}\,\mathbf{u}_2 \\ \lambda_1\mathbf{u}_2{}^{\mathrm{T}}\,\mathbf{u}_1 & \lambda_2\mathbf{u}_2{}^{\mathrm{T}}\,\mathbf{u}_2 \end{bmatrix}.$$

Using (13.6), we obtain

$$\mathbf{u}_1{}^{\mathrm{T}}\,\mathbf{u}_1 = \mathbf{u}_1 \cdot \mathbf{u}_1 = 1, \qquad \mathbf{u}_1{}^{\mathrm{T}}\,\mathbf{u}_2 = \mathbf{u}_1 \cdot \mathbf{u}_2 = 0,$$

and so on.]

15 JACOBIANS

Let R be some region in XYZ-space[1] and let the point P with coordinates (x,y,z) range over all points of R. Suppose that F is a function which assigns to each point P of R some real value $F(x,y,z)$. We shall refer to F as a function of the three variables x, y, z, defined on the region R. *To avoid endless repetition, we assume once and for all that when we write down an expression like $F(x,y,z)$, the point with coordinates (x,y,z) lies in the region R on which the function F is defined.*

Given a function F, we may form its partial derivative with respect to x, denoted by either $\frac{\partial F}{\partial x}$ or F_x. This partial is obtained by differentiating F with respect to x, treating the other variables y, z as constants. For example, if

$$F(x,y,z) = x^2 + y^3z, \tag{15.1}$$

then

$$F_x(x,y,z) = 2x, \qquad F_y(x,y,z) = 3y^2z, \qquad F_z(x,y,z) = y^3.$$

In general, F_x is defined by

$$F_x(x,y,z) = \lim_{h\to 0} \frac{1}{h}\{F(x+h,y,z) - F(x,y,z)\}.$$

This limit may very well fail to exist at certain points of the region R on which F is defined. We shall say that F is *well-behaved* if all of the partial derivatives F_x, F_y, F_z exist and are continuous at every point of R. In order to use certain standard theorems from calculus, such as the chain rule, it will be necessary for us to restrict our attention to well-behaved functions. *We assume once and for all that the functions considered below are well-behaved.*

Suppose now that F is a function of x, y, z, where x, y, z are them-

[1] We shall not give a precise definition of "region"; for example, the reader may visualize R as the set of points inside a sphere.

selves functions of two new variables t, u. As the point Q with coordinates (t,u) ranges over some region R' in the TU-plane, suppose that the corresponding point P with coordinates (x,y,z) always lies in the region R on which F is defined. Then we may view F as a function of two variables t and u, determined by the succession of mappings

$$(t,u) \longrightarrow (x,y,z) \longrightarrow F(x,y,z).$$

This means that the value of F at the point Q is obtained by first finding the point P which corresponds to Q, and then evaluating F at P.

This is not very mysterious in practice. Let F be defined by (15.1), and suppose for example that

$$x = t - u, \qquad y = t^2, \qquad z = u^3.$$

Then F assigns to each point Q in the TU-plane the value

$$\begin{aligned} F(t-u,\, t^2,\, u^3) &= (t-u)^2 + (t^2)^3 u^3 \\ &= t^2 - 2tu + u^2 + t^6 u^3. \end{aligned}$$

In brief, in order to evaluate F at the point Q, just substitute for x, y, z in terms of t and u.

Viewing F as a function of the two variables t, u, it makes sense to form the partial derivative F_t, in which u is held constant. Likewise we can form F_u, keeping t constant. In the above example,

$$\begin{aligned} F_t(t-u,\, t^2,\, u^3) &= \frac{\partial}{\partial t}(t^2 - 2tu + u^2 + t^6 u^3) \\ &= 2t - 2u + 6t^5 u^3. \end{aligned}$$

(15.2) Chain Rule

Let F be a well-behaved function of the three variables x, y, z, each of which is a well-behaved function of two new variables t, u. Then

(15.3)
$$F_t = F_x \cdot \frac{\partial x}{\partial t} + F_y \cdot \frac{\partial y}{\partial t} + F_z \cdot \frac{\partial z}{\partial t}.$$

The proof may be found in most advanced calculus texts, and we do not attempt it here. It may be instructive to rewrite formula (15.3) as

$$\frac{\partial F}{\partial t} = \frac{\partial F}{\partial x} \cdot \frac{\partial x}{\partial t} + \frac{\partial F}{\partial y} \cdot \frac{\partial y}{\partial t} + \frac{\partial F}{\partial z} \cdot \frac{\partial z}{\partial t}.$$

Note that in this formula, we have

$$\frac{\partial F}{\partial t} = \left(\frac{\partial F}{\partial t}\right)_{u=\text{constant}}, \qquad \frac{\partial F}{\partial x} = \left(\frac{\partial F}{\partial x}\right)_{y,z\ \text{constant}},$$

$$\frac{\partial x}{\partial t} = \left(\frac{\partial x}{\partial t}\right)_{u=\text{constant}}, \text{ and so on.}$$

In evaluating F_x, F_y, F_z we need not know anything about how x, y, z are related to t, u; instead, we only need the expression for $F(x,y,z)$ in terms of x, y, z. On the other hand, in calculating $\frac{\partial x}{\partial t}$ we do not use any information about F, but only need the expression for x in terms of t, u.

Let us rewrite (15.3) in the form

$$\left[\frac{\partial F}{\partial t}\right] = [F_x \quad F_y \quad F_z]\begin{bmatrix}\frac{\partial x}{\partial t} \\ \frac{\partial y}{\partial t} \\ \frac{\partial z}{\partial t}\end{bmatrix}.$$

Analogously we obtain

$$\left[\frac{\partial F}{\partial u}\right] = [F_x \quad F_y \quad F_z]\begin{bmatrix}\frac{\partial x}{\partial u} \\ \frac{\partial y}{\partial u} \\ \frac{\partial z}{\partial u}\end{bmatrix}.$$

Combining these into a single matrix equation, we write

$$\text{(15.4)} \qquad \begin{bmatrix}\frac{\partial F}{\partial t} & \frac{\partial F}{\partial u}\end{bmatrix} = [F_x \quad F_y \quad F_z]\begin{bmatrix}\frac{\partial x}{\partial t} & \frac{\partial x}{\partial u} \\ \frac{\partial y}{\partial t} & \frac{\partial y}{\partial u} \\ \frac{\partial z}{\partial t} & \frac{\partial z}{\partial u}\end{bmatrix}.$$

The above discussion suggests that it may be useful to consider matrices whose entries are certain partial derivatives. Of special importance is the *Jacobian matrix*

$$\begin{bmatrix}F_x & F_y & F_z \\ G_x & G_y & G_z \\ H_x & H_y & H_z\end{bmatrix},$$

where F, G, H are functions of x, y, z. The determinant of this matrix is called the *Jacobian* of F, G, H with respect to x, y, z, and is denoted by $\frac{\partial(F,G,H)}{\partial(x,y,z)}$. Thus by definition

$$\text{(15.5)} \qquad \frac{\partial(F,G,H)}{\partial(x,y,z)} = \begin{vmatrix}F_x & F_y & F_z \\ G_x & G_y & G_z \\ H_x & H_y & H_z\end{vmatrix}.$$

EXAMPLES

1. Given $F(x,y,z) = x^2y + 3z$, $G(x,y,z) = xy - z^2$, $H(x,y,z) = x + yz$, we have

$$\frac{\partial(F,G,H)}{\partial(x,y,z)} = \begin{vmatrix} 2xy & x^2 & 3 \\ y & x & -2z \\ 1 & z & y \end{vmatrix}$$

$$= 2x^2y^2 + 3yz - 2x^2z - 3x - x^2y^2 + 4xyz^2.$$

2. Given $x = r\cos\theta$, $y = r\sin\theta$, we obtain

$$\frac{\partial(x,y)}{\partial(r,\theta)} = \begin{vmatrix} \frac{\partial x}{\partial r} & \frac{\partial x}{\partial \theta} \\ \frac{\partial y}{\partial r} & \frac{\partial y}{\partial \theta} \end{vmatrix} = \begin{vmatrix} \cos\theta & -r\sin\theta \\ \sin\theta & r\cos\theta \end{vmatrix}$$

$$= r(\cos^2\theta + \sin^2\theta) = r.$$

(As a matter of fact, we have not been consistent in our choice of notation. The Jacobian $\frac{\partial(F,G,H)}{\partial(x,y,z)}$ is really a function of the three variables x, y, z, whereas the determinant given in Example 1 is the value of the Jacobian at an arbitrary point with coordinates (x,y,z). The notation becomes very cumbersome if we make this distinction each time. Therefore we will often employ the simpler notation used in these examples, inaccurate though it be.)

We are now ready to prove a generalized version of (15.2).

(15.6) Chain Rule for Jacobians

Let F, G be functions of the two variables x, y, and suppose that both x, y are themselves functions of two new variables t, u. Then F, G may be considered as functions of t, u, and we have

(15.7)
$$\frac{\partial(F,G)}{\partial(t,u)} = \frac{\partial(F,G)}{\partial(x,y)} \cdot \frac{\partial(x,y)}{\partial(t,u)}.$$

PROOF: We begin by verifying the matrix equation

(15.8)
$$\begin{bmatrix} F_t & F_u \\ G_t & G_u \end{bmatrix} = \begin{bmatrix} F_x & F_y \\ G_x & G_y \end{bmatrix} \begin{bmatrix} \frac{\partial x}{\partial t} & \frac{\partial x}{\partial u} \\ \frac{\partial y}{\partial t} & \frac{\partial y}{\partial u} \end{bmatrix}.$$

Let us check, for instance, that both sides have the same (2,1)-entry. The (2,1)-entry on the left is G_t. The (2,1)-entry of the product on the right is

$$G_x \cdot \frac{\partial x}{\partial t} + G_y \cdot \frac{\partial y}{\partial t},$$

which equals G_t by the Chain Rule. The other entries can be checked in a similar manner.

Now take determinants of both sides of (15.8), using the fact that the determinant of a product of matrices equals the product of their determinants. Then

$$\begin{vmatrix} F_t & F_u \\ G_t & G_u \end{vmatrix} = \begin{vmatrix} F_x & F_y \\ G_x & G_y \end{vmatrix} \cdot \begin{vmatrix} \frac{\partial x}{\partial t} & \frac{\partial x}{\partial u} \\ \frac{\partial y}{\partial t} & \frac{\partial y}{\partial u} \end{vmatrix},$$

which is exactly the desired relation (15.7).

In a similar way, one obtains

(15.9) $$\frac{\partial(F,G,H)}{\partial(s,t,u)} = \frac{\partial(F,G,H)}{\partial(x,y,z)} \cdot \frac{\partial(x,y,z)}{\partial(s,t,u)},$$

where F, G, H are functions of x, y, z, each of which is in turn a function of s, t, u.

The rest of this section will be devoted to applications of Jacobians. We begin with the study of nonlinear transformations from one coordinate system to another. Since we shall need many of the concepts introduced in Section 8, the reader is advised to review that section briefly before proceeding with this discussion.

Let P denote a point with coordinates (x,y) in the XY-plane. Introduce two new variables u, v, each of which is a function of the two variables x, y. To fix our ideas, suppose that

(15.10) $$u = x^2 + y^2, \qquad v = 2xy.$$

Let Q be the point in the UV-plane with coordinates (u,v). Then equations (15.10) determine a *mapping*, or *transformation*, which assigns to each point P in the XY-plane the corresponding point Q in the UV-plane. If we denote this transformation by the symbol T, we say that T *maps* P onto Q, and write $T: P \to Q$, or sometimes $P \xrightarrow{T} Q$. This particular transformation defined by (15.10) is *not* a linear transformation, however! *Linear* transformations[2] are those defined by a pair of equations

$$u = ax + by, \qquad v = cx + dy, \qquad a, b, c, d \text{ constants.}$$

[2] From the XY-plane into the UV-plane.

If we were to solve equations (15.10) for x, y in terms of u, v, we would find that x, y could not be expressed as single-valued functions of u, v. This is due to the fact that two different points P_1, P_2 in the XY-plane might be mapped onto the *same* point in the UV-plane. For example, let P_1 have coordinates (1,0), and P_2 (0,1); the transformation T defined by (15.10) maps both P_1 and P_2 onto the point with coordinates (1,0) in the UV-plane.

We shall eventually find a region R in the XY-plane such that no two points of R are mapped by T onto the same point in the UV-plane. Supposing we have such a region R, let S be the set of all points Q which correspond to points P of R. Then S is a region in the UV-plane, and to each point P in R there corresponds one and only one point Q in S. Conversely, each point Q in S comes from one and only one point P in R. We shall say that T establishes a *one-to-one correspondence* $P \leftrightarrow Q$ between the points of R and the points of S. (See Figure 15.1.)

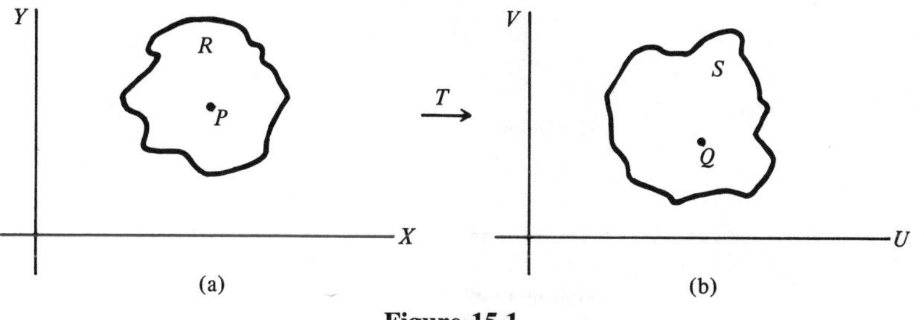

Figure 15.1

Let T' be the reverse mapping, which maps each point Q in S onto its corresponding point P in R; we write T': $Q \to P$. Note that T' is defined only on the region S, and not necessarily on the entire UV-plane. We call T' an *inverse* of T. Symbolically,

$$P \xrightarrow{T} Q \xrightarrow{T'} P, \qquad Q \xrightarrow{T'} P \xrightarrow{T} Q, \qquad P \text{ in } R,\ Q \text{ in } S.$$

We may rephrase the above discussion as follows: there exist functions F^*, G^* defined on the region S, such that the transformation T' is determined by the formulas

$$x = F^*(u,v), \qquad y = G^*(u,v).$$

For each point Q in S with coordinates (u,v), the above equations determine a point P in R with coordinates (x,y), and these variables x, y, u, v satisfy the original equations (15.10).

It is time to consider the key question: how can we find a region R in the XY-plane such that no two points of R are mapped by T onto the same point in the UV-plane? To answer this question, we first evaluate the Jacobian $\frac{\partial(u,v)}{\partial(x,y)}$. We have

$$\frac{\partial(u,v)}{\partial(x,y)} = \begin{vmatrix} \frac{\partial u}{\partial x} & \frac{\partial u}{\partial y} \\ \frac{\partial v}{\partial x} & \frac{\partial v}{\partial y} \end{vmatrix} = \begin{vmatrix} 2x & 2y \\ 2y & 2x \end{vmatrix} = 4(x^2 - y^2).$$

Now let P_0 be any point in the XY-plane at which the Jacobian $\frac{\partial(u,v)}{\partial(x,y)}$ is different from zero. It then turns out (see 15.11) that there necessarily exists some region R surrounding P_0, such that T gives a one-to-one correspondence between the points P of R and the points Q of the corresponding region S in the UV-plane. We shall not enter into the complicated question of finding the largest possible region R with this desired property. In our particular case, P_0 can be any point (x_0,y_0) for which $x_0{}^2 - y_0{}^2 \neq 0$. Choosing P_0 to be the point (1,0), it is not too difficult to prove that R may be taken to be the shaded region shown below, defined by the conditions: $-x < y < x$, $x > 0$. We have also sketched in Figure 15.2 the corresponding region S in the UV-plane, which by a strange coincidence looks like R.

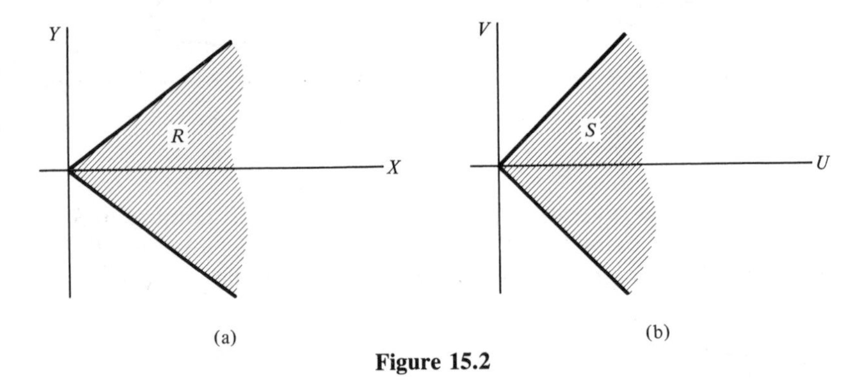

Figure 15.2

The following general result is of basic importance in the theory of functions of several variables.

(15.11) Implicit Function Theorem[3]

Let F, G be a pair of well-behaved functions of two variables x, y, defined on a region R of the XY-plane. Define a transformation T of the region R onto a region S of the UV-plane, by letting T map the point P with coordinates (x,y) onto the point Q in the UV-plane with coordi-

[3] For the proofs of (15.11), (15.17), and (15.19), see any text on advanced calculus or real analysis.

nates (u,v), where

(15.12) $$u = F(x,y), \qquad v = G(x,y).$$

Let P_0 be some point in R at which $\dfrac{\partial(F,G)}{\partial(x,y)}$ is not zero. Then there exists some region N containing P_0, and lying inside R, such that T gives a one-to-one correspondence $P \leftrightarrow Q$ between the points P of N and the points Q of the region N' in the UV-plane which corresponds to N. There is an inverse transformation $T'\colon Q \to P$ mapping N' onto N, given by a pair of formulas

(15.13) $$x = F^*(u,v), \qquad y = G^*(u,v),$$

where F^*, G^* are well-behaved functions defined on N'. Equations (15.12) *and* (15.13) *are consistent with each other in the following sense: If we take the expressions for x, y given in* (15.13), *and substitute them into equations* (15.12), *we will obtain identities valid at each point Q of N'. Likewise, if we substitute in* (15.13) *the expressions for u, v given in* (15.12), *we will obtain identities valid at each point P of N.*

Keeping the above notation, we turn now to the question of how the transformation T affects areas. We shall assume that T gives a one-to-one transformation of the region R onto the region S, and that the Jacobian $\dfrac{\partial(F,G)}{\partial(x,y)}$ is nonzero at every point of R. Let $P_0(x_0,y_0)$ be a point in R, and let $T\colon P_0 \to Q_0$, where Q_0 is the point (u_0,v_0) in S. Suppose that N is some small region in R surrounding P_0; then T maps N onto some region N' in S surrounding Q_0. How does the area of N' compare with the area of N? (See Figure 15.3.)

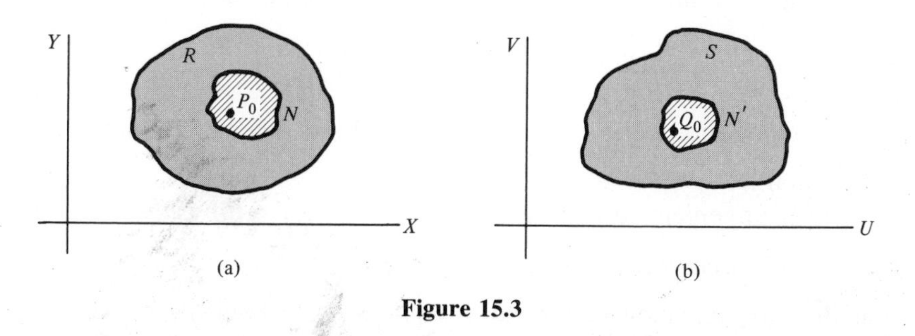

Figure 15.3

We shall give an intuitive discussion of this question, referring the reader to standard texts on advanced calculus or real analysis for detailed proofs. Let P be a point in R near P_0 with coordinates $(x_0 + \Delta x,\, y_0 + \Delta y)$, and let T map P onto the point Q in S with coordinates $(u_0 + \Delta u,\, v_0 + \Delta v)$. Then there are approximation formulas

(15.14)
$$\Delta u \doteq F_x(P_0) \cdot \Delta x + F_y(P_0) \cdot \Delta y,$$
$$\Delta v \doteq G_x(P_0) \cdot \Delta x + G_y(P_0) \cdot \Delta y.$$

Here the symbol $\doteq$ means "is approximately equal to." The notation $F_x(P_0)$ indicates the value of F_x at P_0, that is, $F_x(x_0,y_0)$. The closer Δx and Δy are to zero, the better the approximations.

Equations (15.14) tell us that when P is close to P_0, the transformation defined by equations (15.12) is approximately a linear transformation, with matrix of coefficients

$$\begin{bmatrix} F_x(P_0) & F_y(P_0) \\ G_x(P_0) & G_y(P_0) \end{bmatrix}.$$

We saw in Section 8 that a linear transformation magnifies all areas by a factor equal to the absolute value of the determinant of coefficients. In this case, the determinant is precisely the Jacobian $\dfrac{\partial(F,G)}{\partial(x,y)}$ evaluated at P_0; its absolute value is therefore

(15.15)
$$\left|\frac{\partial(F,G)}{\partial(x,y)}\right|_{P_0}.$$

The preceding intuitive discussion shows that the transformation carrying N onto N' is "approximately" a linear transformation, and so

(15.16)
$$\text{area of } N' \doteq \left|\frac{\partial(F,G)}{\partial(x,y)}\right|_{P_0} \cdot \text{area of } N.$$

Note that since P_0 lies in R, the factor which multiplies the area of N is positive.

We expect that the approximations in (15.14) and (15.16) get better as the region N shrinks around P_0. Indeed, it can be shown that[4]

(15.17)
$$\lim_{N\to 0} \frac{\text{area of } N'}{\text{area of } N} = \left|\frac{\partial(F,G)}{\partial(x,y)}\right|_{P_0}.$$

In calculating $\lim_{N\to 0}$, we let N lie within a circle with center P_0 and radius ϵ, and then let $\epsilon \to 0$.

It is convenient to restate formula (15.17) thus: *the transformation T magnifies areas near P_0 by approximately the factor* $\left|\dfrac{\partial(F,G)}{\partial(x,y)}\right|_{P_0}$. Note that this magnification factor varies from point to point in R. We shall often write $\dfrac{\partial(u,v)}{\partial(x,y)}$ in place of $\dfrac{\partial(F,G)}{\partial(x,y)}$. Further, we may just as well use P, rather than P_0, to denote an arbitrary point of the region R. Rephrasing our results, we have:

[4] See footnote to (15.11).

In passing from the XY-plane to the UV-plane by means of the transformation defined by (15.12), *small areas near a point P are magnified (approximately) by the factor* $\left|\frac{\partial(u,v)}{\partial(x,y)}\right|$ *evaluated at P.*

The preceding result leads us to expect that the area inside S (if finite) can be computed once we know the region R and how the magnification factor $\left|\frac{\partial(u,y)}{\partial(x,y)}\right|$ depends on P. Intuitively, then, we expect that

(15.18) $$\text{area inside } S = \iint_S du\, dv = \iint_R \left|\frac{\partial(u,v)}{\partial(x,y)}\right| dx\, dy.$$

This formula is of great significance, since it suggests how to change variables in multiple integrals. A few minor restrictions are needed to guarantee that (15.18) holds true. The general result is as follows:

(15.19) Theorem[5]

Let T be the transformation from the XY-plane to the UV-plane defined by the equations $u = F(x,y)$, $v = G(x,y)$, *where F, G are well-behaved functions defined on the region R of the XY-plane. Assume that T gives a one-to-one transformation of R onto a region S in the UV-plane, and that* $\frac{\partial(F,G)}{\partial(x,y)} \neq 0$ *at every point of R. Let* R_0 *be some closed bounded region*[6] *inside R, and let T map* R_0 *onto the closed bounded region* S_0 *of the UV-plane.* (See Figure 15.4.) *If* φ *is any function defined and continuous at every point of* S_0*, then*

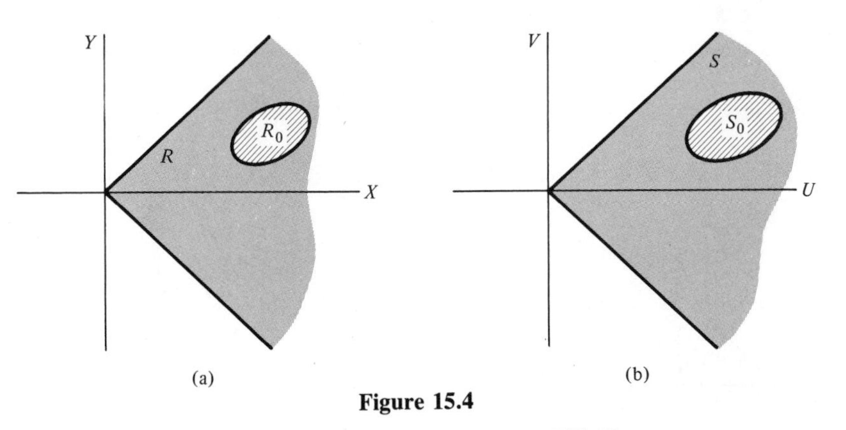

Figure 15.4

[5] As a matter of fact, this theorem remains true when $\frac{\partial(F,G)}{\partial(x,y)} \neq 0$ except for a finite number of points of R; see footnote to (15.11).

[6] We shall not define this term precisely; the reader should visualize R_0 as the set of points inside or on some simple curve.

(15.20) $$\iint_{S_0} \varphi(u,v)\, du\, dv = \iint_{R_0} \varphi(F(x,y),G(x,y)) \cdot \left|\frac{\partial(u,v)}{\partial(x,y)}\right| dx\, dy.$$

In brief, to go from $\iint_{S_0}$ to $\iint_{R_0}$, substitute for u, v in terms of x, y, and replace $du\, dv$ by $\left|\frac{\partial(u,v)}{\partial(x,y)}\right| dx\, dy$.

In the special case where we change from rectangular coordinates (x,y) to polar coordinates (r,θ), we have

$$x = r \cos \theta, \qquad y = r \sin \theta, \qquad \frac{\partial(x,y)}{\partial(r,\theta)} = r.$$

Formula (15.20) then tells us that when we substitute for x, y in terms of r, θ, we must replace $dx\, dy$ by $r\, dr\, d\theta$ (assuming that we keep $r \geqslant 0$). This fact is probably familiar to the reader.

Let us illustrate Theorem 15.19 with an example, to help clarify the discussion.

EXAMPLE

Consider the equations $u = x + y$, $v = y^2$, expressing the variables u, v in terms of x and y. Then

$$\frac{\partial(u,v)}{\partial(x,y)} = \begin{vmatrix} \frac{\partial u}{\partial x} & \frac{\partial u}{\partial y} \\ \frac{\partial v}{\partial x} & \frac{\partial v}{\partial y} \end{vmatrix} = \begin{vmatrix} 1 & 1 \\ 0 & 2y \end{vmatrix} = 2y.$$

Choose R to be the region in the XY-plane consisting of all points (x,y)

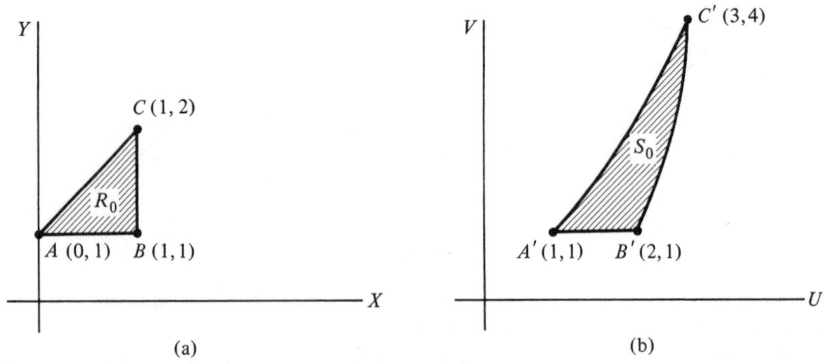

Figure 15.5

for which $y > 0$; thus R is the "upper half-plane." In that case, $\frac{\partial(u,v)}{\partial(x,y)} > 0$ at every point of R. Let T be the transformation carrying (x,y) onto the point (u,v) in the UV-plane. It turns out that T gives a one-to-one mapping of R onto the region S, where S is the upper half of the UV-plane.

Now let A, B, C be points in the XY-plane, with coordinates (0,1), (1,1), (1,2), respectively. Suppose that R_0 is the closed bounded region in the XY-plane consisting of all points on the triangle ABC, together with all points inside that triangle. The transformation T maps R_0 onto the shaded region S_0 in the UV-plane shown in Figure 15.5.

The points on the line segment BC are specified by the conditions $x = 1$, $1 \leqslant y \leqslant 2$. Let $B \xrightarrow{T} B'$, $C \xrightarrow{T} C'$; the transformation T maps the segment BC onto some curve $B'C'$ in the UV-plane. Since $u = x + y$ and $v = y^2$, the parametric equations of $B'C'$ are as follows:

$$\begin{cases} u = 1 + y \\ v = y^2, \qquad\qquad 1 \leqslant y \leqslant 2. \end{cases}$$

Eliminating y, we obtain $v = (u - 1)^2$; thus $B'C'$ is the portion of the parabola $v = (u - 1)^2$ between B' and C'. We may find the remaining boundaries of S_0 in a similar manner, and then S_0 is the closed bounded region consisting of the boundary and interior points shown above.

Now let f be a function defined and continuous at every point of S_0. By (15.20), we obtain

$$\iint_{S_0} f(u,v)\ du\ dv = \iint_{R_0} f(x + y,\ y^2) \cdot |2y|\ dx\ dy.$$

(In fact, we can drop the absolute value sign occurring on the right, since $2y > 0$ throughout R_0.) Thus, for example, we have

$$\iint_{S_0} du\ dv = \iint_{R_0} 2y\ dx\ dy,$$

$$\iint_{S_0} uv\ du\ dv = \iint_{R_0} (x + y) \cdot y^2 \cdot 2y\ dx\ dy,$$

and so on. We shall not evaluate these integrals here.

As another application of Jacobians, we shall consider the problem of computing partial derivatives of implicit functions. Suppose for example that we start with a pair of simultaneous equations

$$\begin{cases} x^2 - y^2 + uv = 7 \\ 2xy - u^2 + 3v^2 = 14. \end{cases}$$

We think of solving these for u, v in terms of x, y. How can we find $\frac{\partial u}{\partial x}$ (keeping y constant), without actually solving explicitly for u, v in terms of x, y? One approach is to use the method of "implicit differentiation," in which we differentiate each of the original equations with respect to x, keeping y constant, while treating u, v as implicit functions of x, y. This gives us the following pair of simultaneous equations:

$$\begin{cases} 2x + u\dfrac{\partial v}{\partial x} + v\dfrac{\partial u}{\partial x} = 0 \\ 2y - 2u\dfrac{\partial u}{\partial x} + 6v\dfrac{\partial v}{\partial x} = 0. \end{cases}$$

These are a pair of *linear* equations in the "unknowns" $\frac{\partial u}{\partial x}$, $\frac{\partial v}{\partial x}$, and we can solve these equations for these unknowns. Rather than complete the calculation, however, we pass at once to the general case.

Let us be given a pair of simultaneous equations

(15.21) $$F(x,y,u,v) = 0, \qquad G(x,y,u,v) = 0,$$

and suppose that they determine u, v as functions of x, y, when the point with coordinates (x,y) is restricted to lie in some suitable region R of the XY-plane. Let us differentiate the identity $F(x,y,u,v) = 0$ with respect to x, keeping y constant, and treating u, v as functions of x, y. By the Chain Rule, we obtain[7]

$$F_x + F_u \cdot \frac{\partial u}{\partial x} + F_v \cdot \frac{\partial v}{\partial x} = 0.$$

Likewise,

$$G_x + G_u \cdot \frac{\partial u}{\partial x} + G_v \cdot \frac{\partial v}{\partial x} = 0.$$

Rewrite these as a pair of simultaneous equations

$$\begin{cases} F_u \cdot \dfrac{\partial u}{\partial x} + F_v \cdot \dfrac{\partial v}{\partial x} = -F_x \\ G_u \cdot \dfrac{\partial u}{\partial x} + G_v \cdot \dfrac{\partial v}{\partial x} = -G_x. \end{cases}$$

Assuming that $\begin{vmatrix} F_u & F_v \\ G_u & G_v \end{vmatrix} \neq 0$, we may use Cramer's Rule to solve for the "unknown" $\frac{\partial u}{\partial x}$. We obtain

$$\frac{\partial u}{\partial x} = \frac{\begin{vmatrix} -F_x & F_v \\ -G_x & G_v \end{vmatrix}}{\begin{vmatrix} F_u & F_v \\ G_u & G_v \end{vmatrix}} = -\frac{\begin{vmatrix} F_x & F_v \\ G_x & G_v \end{vmatrix}}{\begin{vmatrix} F_u & F_v \\ G_u & G_v \end{vmatrix}}.$$

[7] The subscript v denotes the variable v.

We may rewrite this in the form

$$\text{(15.22)} \qquad \left(\frac{\partial u}{\partial x}\right)_{y=\text{constant}} = -\frac{\dfrac{\partial(F,G)}{\partial(x,v)}}{\dfrac{\partial(F,G)}{\partial(u,v)}}.$$

Likewise, we have

$$\left(\frac{\partial u}{\partial y}\right)_{x=\text{constant}} = -\frac{\dfrac{\partial(F,G)}{\partial(y,v)}}{\dfrac{\partial(F,G)}{\partial(u,v)}}.$$

How can we remember such a complicated-looking formula? If we are trying to find $\left(\frac{\partial u}{\partial x}\right)_{y=c}$, we are obviously thinking of u as a function of two variables x, y. For convenience, call x, y the *independent* variables in this case; the remaining variables u, v are then the *dependent* variables. In formula (15.22), the denominator is the Jacobian of F, G with respect to the dependent variables. The numerator is a Jacobian in which one dependent variable has been replaced by an independent variable; indeed, in finding $\left(\frac{\partial u}{\partial x}\right)_{y=c}$, we replaced the dependent variable u by the independent variable x. [This is reasonable, in a sense: the symbol x occurs in the "denominator" on both sides of equation (15.22).]

Let us illustrate with some examples. We assume throughout that we are dealing with well-behaved functions, and that the Jacobians appearing in the denominators are nonzero at any point where we use these formulas.

EXAMPLES

1. Given simultaneous equations

$$F(x,y,z,u,v) = 0, \qquad G(x,y,z,u,v) = 0, \qquad H(x,y,z,u,v) = 0,$$

find $\left(\frac{\partial x}{\partial u}\right)_{v=c}$.

Since we are computing a partial with respect to u, keeping v constant, we must treat u, v as the independent variables. The dependent variables are then the remaining variables x, y, z. The analogue of formula (15.22) then gives

$$\left(\frac{\partial x}{\partial u}\right)_{v=c} = -\frac{\dfrac{\partial(F,G,H)}{\partial(u,y,z)}}{\dfrac{\partial(F,G,H)}{\partial(x,y,z)}}.$$

The denominator is the Jacobian of F, G, H relative to the dependent variables x, y, z. The numerator is a corresponding Jacobian with repect to u, y, z.

Caution: Don't change the order in which you name the variables! It would be wrong to write the numerator as $\dfrac{\partial(F,G,H)}{\partial(y,u,z)}$.

2. Given

$$xy - uvz = 2, \qquad x^2 - u^2v + yz = 4,$$

find $\left(\dfrac{\partial v}{\partial x}\right)_{y,z \text{ constant}}$.

SOLUTION: Let us set

$$F(x,y,z,u,v) = xy - uvz - 2, \qquad G(x,y,z,u,v) = x^2 - u^2v + yz - 4.$$

In this case, x, y, z are the independent variables, u, v the dependent variables. Therefore

$$\left(\frac{\partial v}{\partial x}\right)_{y,z \text{ constant}} = -\frac{\dfrac{\partial(F,G)}{\partial(u,x)}}{\dfrac{\partial(F,G)}{\partial(u,v)}}.$$

Now we have

$$F_x = y, \qquad F_u = -vz, \qquad F_v = -uz,$$
$$G_x = 2x, \qquad G_u = -2uv, \qquad G_v = -u^2.$$

Therefore

$$\frac{\partial(F,G)}{\partial(u,x)} = \begin{vmatrix} F_u & F_x \\ G_u & G_x \end{vmatrix} = \begin{vmatrix} -vz & y \\ -2uv & 2x \end{vmatrix} = 2uvy - 2xvz.$$

Likewise

$$\frac{\partial(F,G)}{\partial(u,v)} = \begin{vmatrix} F_u & F_v \\ G_u & G_v \end{vmatrix} = \begin{vmatrix} -vz & -uz \\ -2uv & -u^2 \end{vmatrix} = -u^2vz.$$

Therefore

$$\left(\frac{\partial v}{\partial x}\right)_{y,z \text{ constant}} = \frac{2uvy - 2xvz}{u^2vz}.$$

3. Given $F(x,y,z) = 0$, find $\left(\dfrac{\partial z}{\partial x}\right)_{y=c}$.

Here x, y are the independent variables, z the dependent variable. Thus

$$\left(\frac{\partial z}{\partial x}\right)_{y=c} = -\frac{\dfrac{\partial F}{\partial x}}{\dfrac{\partial F}{\partial z}} = -\frac{F_x}{F_z}.$$

EXERCISES

1. Given $F(x,y) = x^2 - y^2$, $G(x,y) = 2xy$, find $\dfrac{\partial(F,G)}{\partial(x,y)}$. What is the relationship between $\dfrac{\partial(F,G)}{\partial(x,y)}$ and $\dfrac{\partial(F,G)}{\partial(y,x)}$?
2. Let (ρ,ϕ,θ) be spherical coordinates of the point (x,y,z) in XYZ-space (see Figure 15.6). Show that

$$x = \rho \sin\phi \cos\theta, \qquad y = \rho \sin\phi \sin\theta, \qquad z = \rho \cos\phi.$$

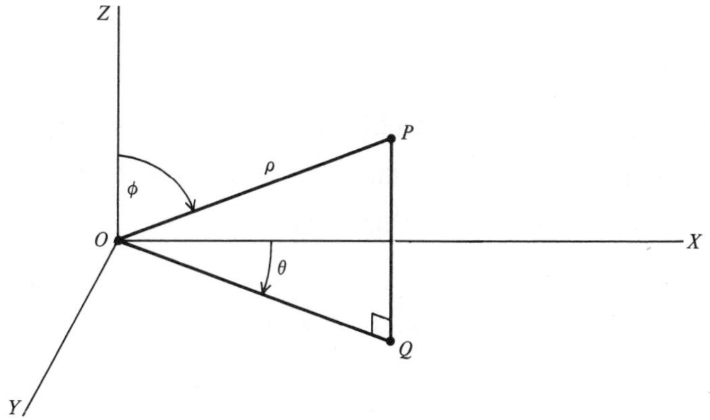

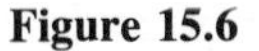

Figure 15.6

Then verify that $\dfrac{\partial(x,y,z)}{\partial(\rho,\phi,\theta)} = \rho^2 \sin\phi$. Show therefore that in evaluating a triple integral $\iiint f(x,y,z)\, dx\, dy\, dz$ by introducing spherical coordinates, the "volume element" $dx\, dy\, dz$ must be replaced by $\rho^2 \sin\phi\, d\rho\, d\phi\, d\theta$.

3. Given $x = r\cos\theta$, and $y = r\sin\theta$, find $\dfrac{\partial(r,\theta)}{\partial(x,y)}$.
[*Hint:* $r = (x^2 + y^2)^{1/2}$, $\theta = \arctan(y/x)$.]
4. Let p, q, r, s be constants, and set

$$x = pu + qv, \qquad y = ru + sv.$$

Given $F(x,y)$ and $G(x,y)$, show that

$$\frac{\partial(F,G)}{\partial(u,v)} = (ps - qr) \cdot \frac{\partial(F,G)}{\partial(x,y)}.$$

$\Big[$*Hint:* By (15.7), $\dfrac{\partial(F,G)}{\partial(u,v)} = \dfrac{\partial(F,G)}{\partial(x,y)} \cdot \dfrac{\partial(x,y)}{\partial(u,v)}$.

But now

$$\frac{\partial(x,y)}{\partial(u,v)} = \begin{vmatrix} \frac{\partial x}{\partial u} & \frac{\partial x}{\partial v} \\ \frac{\partial y}{\partial u} & \frac{\partial y}{\partial v} \end{vmatrix} = \begin{vmatrix} p & q \\ r & s \end{vmatrix} = ps - qr.]$$

5. Given $xy = u + v$, $x^2 - y^2 = 3u - v$, find $\left(\frac{\partial y}{\partial u}\right)_{v=c}$ and $\left(\frac{\partial y}{\partial v}\right)_{u=c}$ by use of Jacobians.
6. Given

$$x^2 + yz = t^2 + u^2, \qquad y^2 - xz = tu, \qquad y + z = t + u,$$

find $\left(\frac{\partial x}{\partial u}\right)_{t=c}$.

7. Let $u = x^2 - y^2$, $v = 2xy$. In Exercise 1, we calculated $\frac{\partial(u,v)}{\partial(x,y)}$. Now use Jacobians to calculate the partials of x, y with respect to u, v, and then evaluate $\frac{\partial(x,y)}{\partial(u,v)}$. Check that

$$\frac{\partial(u,v)}{\partial(x,y)} \cdot \frac{\partial(x,y)}{\partial(u,v)} = 1.$$

8. Let $z = (x^2 + y^2) \cdot e^{xy}$. Find $\left(\frac{\partial x}{\partial z}\right)_{y=\text{constant}}$ and $\left(\frac{\partial x}{\partial y}\right)_{z=\text{constant}}$

*9. Let u, v be functions of x, y, and suppose that x, y are themselves functions of r, s. There is a transformation

$$T_1\colon (r,s) \longrightarrow (x,y)$$

from the RS-plane into the XY-plane, with Jacobian $\frac{\partial(x,y)}{\partial(r,s)}$. Likewise, the transformation

$$T_2\colon (x,y) \longrightarrow (u,v)$$

maps the XY-plane into the UV-plane, and its Jacobian is $\frac{\partial(u,v)}{\partial(x,y)}$. The combination of T_1 followed by T_2 then gives a transformation T_3 from the RS-plane into the UV-plane:

$$(r,s) \xrightarrow{T_1} (x,y) \xrightarrow{T_2} (u,v). \quad (T_3)$$

The Jacobian of T_3 is $\frac{\partial(u,v)}{\partial(r,s)}$. The Chain Rule for Jacobians asserts that

$$\frac{\partial(u,v)}{\partial(r,s)} = \frac{\partial(u,v)}{\partial(x,y)} \cdot \frac{\partial(x,y)}{\partial(r,s)}, \tag{15.23}$$

that is,

$$\text{Jacobian of } T_3 = (\text{Jacobian of } T_2) \cdot (\text{Jacobian of } T_1).$$

Explain the significance of this formula, interpreting absolute values of Jacobians as "magnification factors" for areas.

*10. Keep the notation of the preceding problem, but suppose that T_2 is the inverse of the transformation T_1. Thus

$$(r,s) \xrightarrow{T_1} (x,y) \xrightarrow{T_2} (r,s),$$

$$(x,y) \xrightarrow{T_2} (r,s) \xrightarrow{T_1} (x,y).$$

Show that

$$\frac{\partial(r,s)}{\partial(x,y)} \cdot \frac{\partial(x,y)}{\partial(r,s)} = 1,$$

that is,

(15.24) $$\text{Jacobian of inverse of } T_1 = (\text{Jacobian of } T_1)^{-1}.$$

[*Hint:* In this case we have $u = r$, $v = s$, so $\dfrac{\partial(u,v)}{\partial(r,s)} = 1$ in (15.23).]

*11. Explain the significance of (15.24) in terms of magnification of areas.

12. Let T be the transformation from the XY-plane to the UV-plane defined by

$$u = 2x + y, \qquad v = 3y.$$

Let R_0 consist of all points (x,y) such that $0 \leqslant x \leqslant 1$, $0 \leqslant y \leqslant 1$, and suppose that T maps R_0 onto S_0. Find S_0 explicitly and express

$$\iint_{S_0} (u + 4v)\, du\, dv$$

as a double integral over R_0.

13. Keeping the notation of Problem 12, express $\displaystyle\iint_{R_0} (2x - y)\, dx\, dy$ as a double integral over S_0.

14. Let T be defined by the equations

$$u = x^2 + y, \qquad v = -x + y^2.$$

Let R_0 be the closed region defined by $0 \leqslant y \leqslant x$, $0 \leqslant x \leqslant 1$, and let T map R_0 onto S_0. Determine S_0 explicitly, and express $\displaystyle\iint_{S_0} (u^2 + v^2)\, du\, dv$ as a double integral over R_0.

SOLUTIONS OF SELECTED EXERCISES

(Exercise 2.4 refers to Exercise 4 at the end of Section 2.)

Exercise 2.4 $\begin{bmatrix} 2 & 3 & 4 & 5 \\ 4 & 5 & 6 & 7 \\ 6 & 7 & 8 & 9 \end{bmatrix}$.

Exercise 2.5 $\begin{bmatrix} 1 & 0 & 0 & 0 & 0 \\ 0 & 1 & 0 & 0 & 0 \\ 0 & 0 & 1 & 0 & 0 \\ 0 & 0 & 0 & 1 & 0 \end{bmatrix}$.

Exercise 3.3 The jth row of A^{T} is the jth column of A. Also, the jth column of $(A^{\mathrm{T}})^{\mathrm{T}}$ is the jth row of A^{T}. Thus for each j, A and $(A^{\mathrm{T}})^{\mathrm{T}}$ have the same jth column. Therefore $A = (A^{\mathrm{T}})^{\mathrm{T}}$.

Exercise 3.5 The ith row of A^{T} is $[a_{1i} \quad a_{2i} \quad \cdots \quad a_{mi}]$. The jth column of A^{T} is $\begin{bmatrix} a_{j1} \\ a_{j2} \\ \cdots \\ a_{jn} \end{bmatrix}$.

Exercise 3.6 Let $\mathbf{r}_i = [a_{i1} \quad a_{i2} \quad \cdots \quad a_{in}] = i$th row of A. Then

$$\mathbf{r}_i^{\mathrm{T}} = \begin{bmatrix} a_{i1} \\ a_{i2} \\ \cdot \\ \cdot \\ \cdot \\ a_{in} \end{bmatrix}.$$

Therefore

$$[\mathbf{r}_1^{\mathrm{T}} \quad \cdots \quad \mathbf{r}_m^{\mathrm{T}}] = \begin{bmatrix} a_{11} & \cdots & a_{m1} \\ a_{12} & \cdots & a_{m2} \\ & \cdots & \\ a_{1n} & \cdots & a_{mn} \end{bmatrix} = A^{\mathrm{T}}.$$

Exercise 3.7

$$[1 \quad 0 \quad \cdots \quad 0]\begin{bmatrix} a_{11} & \cdots & a_{1n} \\ a_{21} & \cdots & a_{2n} \\ & \cdots & \\ a_{m1} & \cdots & a_{mn} \end{bmatrix}$$

$$= [1 \cdot a_{11} + 0 \cdot a_{21} + \cdots + 0 \cdot a_{m1}, \ \ldots, \ 1 \cdot a_{1n} + 0 \cdot a_{2n} + \cdots + 0 \cdot a_{mn}]$$

$$= [a_{11} \quad \cdots \quad a_{1n}].$$

In general,

$[0 \quad \cdots \quad 0 \quad 1 \quad 0 \quad \cdots \quad 0]A = i$th row of A, (where 1 is in ith place).

$$A\begin{bmatrix} 0 \\ \cdot \\ \cdot \\ \cdot \\ 1 \\ \cdot \\ \cdot \\ \cdot \\ 0 \end{bmatrix} = j\text{th column of } A, \text{ (where 1 is in } j\text{th place).}$$

Exercise 4.5 If $AB = BA$, then

$$\begin{aligned} AB^2 &= A(BB) = (AB)B = (BA)B \\ &= B(AB) = B(BA) = B^2A. \end{aligned}$$

Exercise 4.6 For a positive integer k, we have

$$(\underbrace{A \cdot A \quad \cdots \quad A}_{k \text{ factors}})^{\mathrm{T}} = \underbrace{A^{\mathrm{T}} \cdot A^{\mathrm{T}} \quad \cdots \quad A^{\mathrm{T}}}_{k \text{ factors}} \quad \text{by repeated use of (4.3).}$$

$$= (A^{\mathrm{T}})^k.$$

Exercise 4.7 Suppose that $AB = BA$ and $AC = CA$. Then

$$AB^l = A(\underbrace{B \cdot B \quad \cdots \quad B}_{l \text{ factors}}) = BA(\underbrace{B \quad \cdots \quad B}_{l-1 \text{ factors}}) = \cdots$$

$$= (\underbrace{B \cdot B \quad \cdots \quad B}_{l \text{ factors}})A = B^lA.$$

Therefore

$$A^kB^l = (\underbrace{A \quad \cdots \quad A}_{k \text{ factors}})B^l = (\underbrace{A \quad \cdots \quad A}_{k-1 \text{ factors}})B^lA = \cdots$$

$$= B^l(\underbrace{A \quad \cdots \quad A}_{k \text{ factors}}) = B^lA^k.$$

Also

$$A(B+C) = AB + AC = BA + CA = (B+C)A.$$

Exercise 4.8 $XAX^{\mathrm{T}} = ax^2 + dy^2 + fz^2 + 2bxy + 2cxz + 2eyz,$
$XX^{\mathrm{T}} = x^2 + y^2 + z^2.$

Exercise 4.10 If $A^{\mathrm{T}} = -A$, then the (i,j)-entry of $A^{\mathrm{T}} = -(i,j)$-entry of A, so $a_{ji} = -a_{ij}$. Hence $a_{ii} = -a_{ii}$, so $a_{ii} = 0$. General skew-symmetric matrices:

$$\begin{bmatrix} 0 & a \\ -a & 0 \end{bmatrix}, \begin{bmatrix} 0 & b & c \\ -b & 0 & d \\ -c & -d & 0 \end{bmatrix}.$$

Exercise 4.11 To show that $B + B^{\mathrm{T}}$ is symmetric, we observe that

$$(B + B^{\mathrm{T}})^{\mathrm{T}} = B^{\mathrm{T}} + (B^{\mathrm{T}})^{\mathrm{T}} = B^{\mathrm{T}} + B = B + B^{\mathrm{T}}.$$

Also, $B - B^{\mathrm{T}}$ is skew-symmetric since

$$(B - B^{\mathrm{T}})^{\mathrm{T}} = B^{\mathrm{T}} - (B^{\mathrm{T}})^{\mathrm{T}} = B^{\mathrm{T}} - B = -(B - B^{\mathrm{T}}).$$

Exercise 4.14 Let

$$\mathbf{v} = [c_1 \quad c_2 \quad \cdots \quad c_n], \qquad \bar{\mathbf{v}} = [\bar{c}_1 \quad \bar{c}_2 \quad \cdots \quad \bar{c}_n].$$

Then

$$\bar{\mathbf{v}}\mathbf{v}^{\mathrm{T}} = [\bar{c}_1 \quad \bar{c}_2 \quad \cdots \quad \bar{c}_n] \begin{bmatrix} c_1 \\ c_2 \\ \cdot \\ \cdot \\ \cdot \\ c_n \end{bmatrix} = \bar{c}_1c_1 + \bar{c}_2c_2 + \cdots + \bar{c}_nc_n$$

$$= |c_1|^2 + |c_2|^2 + \cdots + |c_n|^2.$$

Exercise 5.1

$$\det A = 1 \cdot \begin{vmatrix} 5 & -1 \\ 2 & 3 \end{vmatrix} - 2 \cdot \begin{vmatrix} 4 & -1 \\ 0 & 3 \end{vmatrix} + 3 \begin{vmatrix} 4 & 5 \\ 0 & 2 \end{vmatrix} = 17 - 24 + 24 = 17.$$

Exercise 5.3 Subtract α times row 2 from row 1, and β times row 3 from row 1. This procedure does not change the determinant, and yields a matrix with first row zero.

Exercise 5.7

$$\det A = b_{11} \cdot \begin{vmatrix} b_{22} & 0 & 0 & 0 \\ 0 & c_{11} & c_{12} & c_{13} \\ 0 & c_{21} & c_{22} & c_{23} \\ 0 & c_{31} & c_{32} & c_{33} \end{vmatrix} - b_{12} \begin{vmatrix} b_{21} & 0 & 0 & 0 \\ 0 & c_{11} & c_{12} & c_{13} \\ 0 & c_{21} & c_{22} & c_{23} \\ 0 & c_{31} & c_{32} & c_{33} \end{vmatrix}$$

$$= (b_{11}b_{22} - b_{12}b_{21}) \cdot \det C = (\det B)(\det C).$$

Exercise 5.10 The equation can be written as

$$(y_1 - y_2)x + (x_2 - x_1)y + (x_1y_2 - x_2y_1) = 0.$$

Since $P_1 \neq P_2$, either $x_1 \neq x_2$ or $y_1 \neq y_2$; therefore either the coefficient of y is nonzero, or the coefficient of x is nonzero. The equation is satisfied by the coordinates of P_1, since substituting x_1 for x, and y_1 for y, in the left-hand side of the equation, the left-hand expression becomes

$$\begin{vmatrix} x_1 & y_1 & 1 \\ x_1 & y_1 & 1 \\ x_2 & y_2 & 1 \end{vmatrix},$$

which is surely zero. Likewise, the coordinates of P_2 satisfy the equation.

Exercise 5.12

$$\det (\underbrace{A \cdots A}_{k \text{ factors}}) = \underbrace{(\det A) \cdots (\det A)}_{k \text{ factors}} \text{ by repeated use of (5.5).}$$

Exercise 5.13 If $AB = I$, then $(\det A)(\det B) = \det (AB) = \det I = 1$, so $\det A \neq 0$.

Exercise 6.1

$$[5]^{-1} = [\tfrac{1}{5}].$$

$$\begin{bmatrix} 2 & -1 \\ 3 & 2 \end{bmatrix}^{-1} = \frac{1}{7}\begin{bmatrix} 2 & 1 \\ -3 & 2 \end{bmatrix} = \begin{bmatrix} \frac{2}{7} & \frac{1}{7} \\ \frac{-3}{7} & \frac{2}{7} \end{bmatrix}.$$

$$\begin{bmatrix} 1 & x \\ 0 & 1 \end{bmatrix}^{-1} = \begin{bmatrix} 1 & -x \\ 0 & 1 \end{bmatrix}, \begin{bmatrix} 1 & 0 \\ y & 1 \end{bmatrix}^{-1} = \begin{bmatrix} 1 & 0 \\ -y & 1 \end{bmatrix}.$$

Exercise 6.2 For the matrix $\begin{bmatrix} 2 & 3 \\ -1 & 4 \end{bmatrix}$, the cofactor of 2 is 4, cofactor of 3 is $-(-1)$, cofactor of -1 is -3, cofactor of 4 is 2. The matrix of cofactors is $\begin{bmatrix} 4 & 1 \\ -3 & 2 \end{bmatrix}$, and

$$\operatorname{adj}\begin{bmatrix} 2 & 3 \\ -1 & 4 \end{bmatrix} = \text{transpose of matrix of cofactors} = \begin{bmatrix} 4 & -3 \\ 1 & 2 \end{bmatrix}.$$

For the matrix $\begin{bmatrix} 6 & 1 & 0 \\ 2 & -1 & 1 \\ 0 & 2 & 3 \end{bmatrix}$, the matrix of cofactors equals

$$\begin{bmatrix} \begin{vmatrix} -1 & 1 \\ 2 & 3 \end{vmatrix} & -\begin{vmatrix} 2 & 1 \\ 0 & 3 \end{vmatrix} & \begin{vmatrix} 2 & -1 \\ 0 & 2 \end{vmatrix} \\ -\begin{vmatrix} 1 & 0 \\ 2 & 3 \end{vmatrix} & \begin{vmatrix} 6 & 0 \\ 0 & 3 \end{vmatrix} & -\begin{vmatrix} 6 & 1 \\ 0 & 2 \end{vmatrix} \\ \begin{vmatrix} 1 & 0 \\ -1 & 1 \end{vmatrix} & -\begin{vmatrix} 6 & 0 \\ 2 & 1 \end{vmatrix} & \begin{vmatrix} 6 & 1 \\ 2 & -1 \end{vmatrix} \end{bmatrix} = \begin{bmatrix} -5 & -6 & 4 \\ -3 & 18 & -12 \\ 1 & -6 & -8 \end{bmatrix},$$

so

$$\operatorname{adj}\begin{bmatrix} 6 & 1 & 0 \\ 2 & -1 & 1 \\ 0 & 2 & 3 \end{bmatrix} = \begin{bmatrix} -5 & -3 & 1 \\ -6 & 18 & -6 \\ 4 & -12 & -8 \end{bmatrix}.$$

Exercise 6.12 If A and B are nonsingular $n \times n$ matrices, then AB is also nonsingular because $\det(AB) = (\det A)(\det B) \neq 0$. However, $A + B$ may be singular; for example, if $A = I$, $B = -I$, then A and B are nonsingular, but $A + B = 0$.

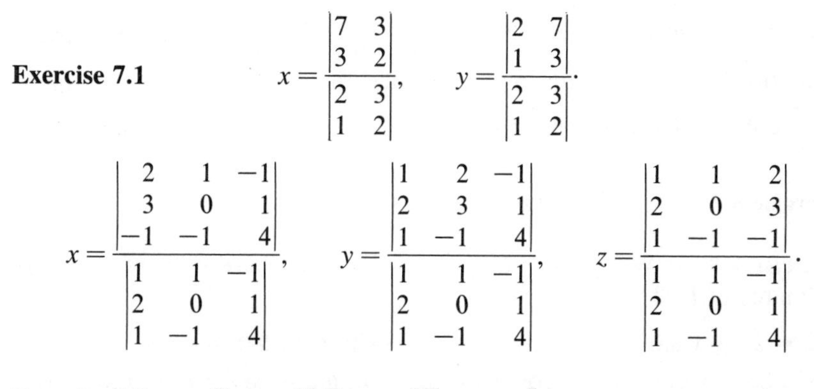

Exercise 7.1

$$x = \frac{\begin{vmatrix} 7 & 3 \\ 3 & 2 \end{vmatrix}}{\begin{vmatrix} 2 & 3 \\ 1 & 2 \end{vmatrix}}, \quad y = \frac{\begin{vmatrix} 2 & 7 \\ 1 & 3 \end{vmatrix}}{\begin{vmatrix} 2 & 3 \\ 1 & 2 \end{vmatrix}}.$$

$$x = \frac{\begin{vmatrix} 2 & 1 & -1 \\ 3 & 0 & 1 \\ -1 & -1 & 4 \end{vmatrix}}{\begin{vmatrix} 1 & 1 & -1 \\ 2 & 0 & 1 \\ 1 & -1 & 4 \end{vmatrix}}, \quad y = \frac{\begin{vmatrix} 1 & 2 & -1 \\ 2 & 3 & 1 \\ 1 & -1 & 4 \end{vmatrix}}{\begin{vmatrix} 1 & 1 & -1 \\ 2 & 0 & 1 \\ 1 & -1 & 4 \end{vmatrix}}, \quad z = \frac{\begin{vmatrix} 1 & 1 & 2 \\ 2 & 0 & 3 \\ 1 & -1 & -1 \end{vmatrix}}{\begin{vmatrix} 1 & 1 & -1 \\ 2 & 0 & 1 \\ 1 & -1 & 4 \end{vmatrix}}.$$

Exercise 7.2 (i) $x = 13/9$, $y = 5/3$, $z = -5$.
(ii) $z = -1$, y arbitrary, $x = 2 - y$.
(iii) z arbitrary, $y = -1 - z$, $x = 2$.

(iv) $x = 4$, $y = 3$, $z = 5$.
(v) inconsistent.
(vi) y arbitrary, z arbitrary, $x = -4 - y - z$.

Exercise 7.4 By Cramer's Rule,

$$x = \frac{\begin{vmatrix} u & b \\ v & d \end{vmatrix}}{\begin{vmatrix} a & b \\ c & d \end{vmatrix}}, \qquad y = \frac{\begin{vmatrix} a & u \\ c & v \end{vmatrix}}{\begin{vmatrix} a & b \\ c & d \end{vmatrix}}.$$

Exercise 7.8 Use Theorem 7.13.

Exercise 8.2 The line $x + y = 0$ in the XY-plane maps onto (0,0) in the UV-plane.

Exercise 8.3 Since $ad - bc \neq 0$, we may use Exercise 7.4 to solve for x, y in terms of u, v. Thus $Ax + By + C = 0$ maps onto the line

$$\frac{1}{ad - bc}\{A(du - bv) + B(-cu + av)\} + C = 0$$

in the UV-plane.

Exercise 8.4 After stretching horizontal distances by the factor 4, and vertical distances by the factor 2, each point is moved to the right 5 units and up 1 unit.

Exercise 8.6 The region inside the sphere $x^2 + y^2 + z^2 = 1$ is mapped onto the region inside the ellipsoid $\frac{u^2}{a^2} + \frac{v^2}{b^2} + \frac{w^2}{c^2} = 1$ by the linear transformation $u = ax$, $v = by$, $w = cz$. Since the volume inside the sphere is $\frac{4\pi}{3}$ cubic units, the volume inside the ellipsoid is $\frac{4\pi}{3}\,abc$ cubic units.

Exercise 8.7 $(x,y) \longrightarrow (x,-y)$.

Exercise 8.9 Write $x = r\cos\theta$, $y = r\sin\theta$, where r, θ are polar coordinates of P. Then

$$x\cos 2\alpha + y\sin 2\alpha = r(\cos\theta\cos 2\alpha + \sin\theta\sin 2\alpha) = r\cos(2\alpha - \theta),$$
$$x\sin 2\alpha - y\cos 2\alpha = r(\cos\theta\sin 2\alpha - \sin\theta\cos 2\alpha) = r\sin(2\alpha - \theta).$$

Thus P maps onto the point P' with polar coordinates r, $2\alpha - \theta$. Since

$$\tfrac{1}{2}(\theta + (2\alpha - \theta)) = \alpha,$$

the points P, P' are symmetrically located with respect to the line L.

Exercise 9.1 Since $\mathbf{v}_3 = \mathbf{v}_1 - \mathbf{v}_2$, the space is spanned by $\{\mathbf{v}_1, \mathbf{v}_2\}$. Neither of the vectors $\{\mathbf{v}_1, \mathbf{v}_2\}$ is a scalar multiple of the other; hence $\{\mathbf{v}_1, \mathbf{v}_2\}$ is a basis for the given vector space.

Exercise 9.2 (These are not the only possible answers.)

Matrix	*Basis for Row Space*	*Basis for Column Space*
$[0 \quad 1]$	$[0 \quad 1]$	$[1]$
$[1 \quad 1]$	$[1 \quad 1]$	$[1]$
$\begin{bmatrix} 2 & 1 \\ 1 & 2 \end{bmatrix}$	$[2 \quad 1], [1 \quad 2]$	$\begin{bmatrix} 2 \\ 1 \end{bmatrix}, \begin{bmatrix} 1 \\ 2 \end{bmatrix}$
$\begin{bmatrix} 1 & 0 & 0 & 1 \\ 0 & 0 & 0 & 1 \\ 0 & 1 & 0 & 1 \end{bmatrix}$	$[1 \quad 0 \quad 0 \quad 1], [0 \quad 0 \quad 0 \quad 1], [0 \quad 1 \quad 0 \quad 1]$	$\begin{bmatrix} 1 \\ 0 \\ 0 \end{bmatrix}, \begin{bmatrix} 0 \\ 0 \\ 1 \end{bmatrix}, \begin{bmatrix} 1 \\ 1 \\ 1 \end{bmatrix}$
$\begin{bmatrix} 1 & 0 & 1 \\ 0 & 1 & 1 \\ 1 & 0 & -1 \\ 0 & 1 & -1 \end{bmatrix}$	$[1 \quad 0 \quad 1], [0 \quad 1 \quad 1], [0 \quad 0 \quad -2]$	$\begin{bmatrix} 1 \\ 0 \\ 1 \\ 0 \end{bmatrix}, \begin{bmatrix} 0 \\ 1 \\ 0 \\ 1 \end{bmatrix}, \begin{bmatrix} 0 \\ 0 \\ -2 \\ -2 \end{bmatrix}$
$\begin{bmatrix} 0 & 1 & 0 \\ 0 & 3 & 0 \end{bmatrix}$	$[0 \quad 1 \quad 0]$	$\begin{bmatrix} 1 \\ 3 \end{bmatrix}$

Exercise 9.3 Nonzero minors

$$|1|, \quad |1|, \quad \begin{vmatrix} 2 & 1 \\ 1 & 2 \end{vmatrix}, \quad \begin{vmatrix} 1 & 0 & 1 \\ 0 & 0 & 1 \\ 0 & 1 & 1 \end{vmatrix}, \quad \begin{vmatrix} 1 & 0 & 1 \\ 0 & 1 & 1 \\ 1 & 0 & -1 \end{vmatrix}, \quad |1|.$$

Exercise 9.5 Let C be a submatrix of A, and let $r =$ rank of A, $s =$ rank of C. By Theorem 9.4(iv), some $s \times s$ minor of C is not zero. This minor is also an $s \times s$ minor of A. But if $s > r$, then by Theorem 9.4(iv), every $s \times s$ minor of A would be zero. Therefore we must have $s \leqslant r$.

Exercise 9.6 The first $n - 1$ rows of A are linearly independent, and the last row is the zero vector. Thus A has rank $n - 1$. Further, A^2 has zeros everywhere except for 1's along the line parallel to the main diagonal, starting at the (1,3) position. In other words, to find A^2 we shift each 1 in A one place to the right. Likewise, A^3 is gotten from A by shifting each 1 in A two places to the right. Eventually one gets $A^n = 0$, and

$$\text{rank } A^2 = n - 2, \text{ rank } A^3 = n - 3, \ \ldots \ , \text{rank } A^{n-1} = 1, \text{ rank } A^n = 0.$$

Exercise 9.8 Rank equals 1 if and only if three planes coincide. Rank equals 2 if there are at least two distinct planes and the third plane passes through their line of intersection. Rank equals 3 if there are three distinct planes whose only common point is the origin.

Exercise 9.10 Let A be the $k \times n$ matrix with rows $\mathbf{r}_1, \ldots, \mathbf{r}_k$. By definition, the rank of A is the dimension of the vector space spanned by the row vectors $\mathbf{r}_1, \ldots, \mathbf{r}_k$. This dimension equals k if the set of vectors $\{\mathbf{r}_1, \ldots, \mathbf{r}_k\}$ is linearly independent; if not, the dimension is less than k. Hence the rank of A equals k if and only if $\{\mathbf{r}_1, \ldots, \mathbf{r}_k\}$ is a linearly independent set of vectors.

Exercise 9.11 Let $\mathbf{r}_1, \ldots, \mathbf{r}_{n+1}$ be a set of $1 \times n$ vectors, and let A be the $(n+1) \times n$ matrix with rows $\mathbf{r}_1, \ldots, \mathbf{r}_{n+1}$. By Theorem 9.4(i), rank of $A \leqslant$ number of columns of A. Thus rank of $A \leqslant n$. By Exercise 9.10 (with $k = n + 1$) it follows at once that the set of vectors $\{\mathbf{r}_1, \ldots, \mathbf{r}_{n+1}\}$ is *not* linearly independent.

Exercise 9.12 Suppose $A^{n \times n}$ is singular. Then $\det A = 0$, and so also $\det A^{\mathrm{T}} = 0$. By Theorem 7.13, there exists a nonzero vector $\begin{bmatrix} c_1 \\ \cdot \\ \cdot \\ \cdot \\ c_n \end{bmatrix}$ such that $A^{\mathrm{T}} \begin{bmatrix} c_1 \\ \cdot \\ \cdot \\ \cdot \\ c_n \end{bmatrix} = \mathbf{0}$. Taking transposes, we get

$$[c_1 \quad \cdots \quad c_n]A = \mathbf{0}.$$

But this means that

(9.12a)
$$c_1\mathbf{r}_1 + \cdots + c_n\mathbf{r}_n = \mathbf{0},$$

where $\mathbf{r}_1, \ldots, \mathbf{r}_n$ are the rows of A. The argument can be reversed to show that if there exist constants $c_1, \ldots, c_n$ not all zero such that (9.12a) holds, then A must be singular.

Exercise 9.13 If $\alpha_1\mathbf{v}_1 + \cdots + \alpha_k\mathbf{v}_k = \beta_1\mathbf{v}_1 + \cdots + \beta_k\mathbf{v}_k$, then

$$(\alpha_1 - \beta_1)\mathbf{v}_1 + \cdots + (\alpha_k - \beta_k)\mathbf{v}_k = \mathbf{0}.$$

If (say) $\alpha_1 \neq \beta_1$, we could solve the above equation for $\mathbf{v}_1$ as a linear combination of $\mathbf{v}_2, \ldots, \mathbf{v}_k$:

$$\mathbf{v}_1 = -\left(\frac{\alpha_2 - \beta_2}{\alpha_1 - \beta_1}\right)\mathbf{v}_2 - \cdots - \left(\frac{\alpha_k - \beta_k}{\alpha_1 - \beta_1}\right)\mathbf{v}_k.$$

But this is impossible since the vectors $\{\mathbf{v}_1, \ldots, \mathbf{v}_k\}$ form a linearly independent set. Thus we must have $\alpha_1 = \beta_1, \ldots, \alpha_k = \beta_k$.

Since $\{\mathbf{v}_1, \ldots, \mathbf{v}_k\}$ span V, every vector $\mathbf{v}$ in V is expressible as linear combination of $\mathbf{v}_1, \ldots, \mathbf{v}_k$. The preceding argument shows that $\mathbf{v}$ cannot be expressed as two different such linear combinations.

Exercise 9.14 The discussion at the beginning of Section 9 shows that these two sets of vectors both span V. Let

$$\mathbf{w} = \mathbf{v}_1 + c_2\mathbf{v}_2 + \cdots + c_k\mathbf{v}_k.$$

If $\{\mathbf{w}, \mathbf{v}_2, \ldots, \mathbf{v}_k\}$ is not a basis of V, then one of these vectors is expressible as a linear combination of the others. Surely $\mathbf{w}$ is not a linear combination of $\mathbf{v}_2, \ldots, \mathbf{v}_k$; for if it were, we could solve for $\mathbf{v}_1$ as a linear combination of $\mathbf{v}_2, \ldots, \mathbf{v}_k$ (which is impossible). On the other hand, suppose that (say)

$$\mathbf{v}_2 = \alpha\mathbf{w} + \alpha_3\mathbf{v}_3 + \cdots + \alpha_k\mathbf{v}_k.$$

If $\alpha \neq 0$, we could solve for $\mathbf{w}$ in terms of $\mathbf{v}_2, \mathbf{v}_3, \ldots, \mathbf{v}_k$, and we just saw that this cannot happen. If $\alpha = 0$, then $\mathbf{v}_2$ is expressed in terms of $\mathbf{v}_3, \ldots, \mathbf{v}_k$, which is also impossible. Therefore no one of the vectors $\{\mathbf{w}, \mathbf{v}_2, \ldots, \mathbf{v}_k\}$ is expressible in terms of the others; hence these vectors form a basis for V.

Exercise 9.15 Let $r =$ rank of A, $s =$ rank of B. If (say) every row of A is expressible as a linear combination of the first r rows of A, and every row of B in terms of the first s rows of B, then every row of $\begin{bmatrix} A & 0 \\ 0 & B \end{bmatrix}$ is expressible in terms of rows number $1, \ldots, r, m+1, \ldots, m+s$. These rows are independent, so rank $\begin{bmatrix} A & 0 \\ 0 & B \end{bmatrix}$ equals $r + s$.

Exercise 9.17 This is a restatement of the discussion preceding (7.12). It also follows from Theorem 9.7, using $\mathbf{b} = \mathbf{0}$.

Exercise 9.18 Immediate from Theorem 9.3.

Exercise 9.19 The system of equations

$$[y_1 \ \cdots \ y_m]A = [0 \ \cdots \ 0]$$

has the same solutions as the system

$$A^{\mathrm{T}} \begin{bmatrix} y_1 \\ \cdot \\ \cdot \\ \cdot \\ y_m \end{bmatrix} = \begin{bmatrix} 0 \\ \cdot \\ \cdot \\ \cdot \\ 0 \end{bmatrix}.$$

Since rank $A^{\mathrm{T}} = \text{rank } A = r$, we can solve for r of the y's in terms of the remaining $m - r$ y's, which may have arbitrary values.

Exercise 10.1 The transformation is given by $\mathbf{v} \to A\mathbf{v}$, where

$$A\begin{bmatrix}1\\0\end{bmatrix} = \begin{bmatrix}3\\1\end{bmatrix}, \qquad A\begin{bmatrix}0\\1\end{bmatrix} = \begin{bmatrix}-1\\1\end{bmatrix}.$$

Thus

$$A = \begin{bmatrix}3 & -1\\1 & 1\end{bmatrix}.$$

Exercise 10.3 We must find $A^{3\times 3}$ such that

$$A \cdot \begin{bmatrix}-1 & 0 & 0\\1 & 1 & 4\\1 & 2 & 3\end{bmatrix} = \begin{bmatrix}2 & 3 & 1\\1 & 2 & 1\\4 & 0 & 1\end{bmatrix}.$$

This gives

$$A = \begin{bmatrix}2 & 3 & 1\\1 & 2 & 1\\4 & 0 & 1\end{bmatrix} \cdot \begin{bmatrix}-1 & 0 & 0\\1 & 1 & 4\\1 & 2 & 3\end{bmatrix}^{-1}.$$

The desired transformation is given by $\mathbf{v} \to A\mathbf{v}$.

Exercise 10.5 Kernel consists of all column vectors $[-2y - z - w, y, z, w, 3y - z + w]^{\mathrm{T}}$, with y, z, w arbitrary. Range is the vector space spanned by $\begin{bmatrix}1\\0\end{bmatrix}$ and $\begin{bmatrix}0\\1\end{bmatrix}$.

Exercise 10.6 Write $\mathbf{v} = \alpha_1\mathbf{v}_1 + \cdots + \alpha_k\mathbf{v}_k$. Then

$$F(\mathbf{v}) = \alpha_1 F(\mathbf{v}_1) + \cdots + \alpha_k F(\mathbf{v}_k).$$

Exercise 10.8 If $A\mathbf{v} = \mathbf{0}$, also $(BA)\mathbf{v} = \mathbf{0}$. Hence every vector in the null space of A is also in the null space of BA. Conversely, if $(BA)\mathbf{w} = \mathbf{0}$, then $B^{-1} \cdot (BA)\mathbf{w} = \mathbf{0}$, that is, $A\mathbf{w} = \mathbf{0}$. Hence every vector in the null space of BA is also in the null space of A.

Let s = rank of null space of A = rank of null space of BA. By Theorem 10.6,

$$\text{rank of } A = n - (\text{rank of null space of } A) = n - s$$
$$\text{rank of } BA = n - (\text{rank of null space of } BA) = n - s.$$

Hence rank of A = rank of BA.

Exercise 11.2 The equations may be written as

$$A\mathbf{v} = c\mathbf{v}, \qquad A = \begin{bmatrix}2 & 1 & -1\\0 & 1 & -1\\0 & 8 & -5\end{bmatrix}, \qquad \mathbf{v} = \begin{bmatrix}x\\y\\z\end{bmatrix}.$$

Then c must be a characteristic root of A. The characteristic equation of A is

$$\begin{vmatrix} \lambda-2 & -1 & 1 \\ 0 & \lambda-1 & 1 \\ 0 & -8 & \lambda+5 \end{vmatrix} = (\lambda-2)(\lambda^2+4\lambda+3)=0,$$

with roots $\lambda = 2, -1, -3$. For each such λ, solve $(\lambda I - A)\mathbf{v} = \mathbf{0}$ to find the possible values of x,y,z.

Exercise 11.4 Use the fact that

$$(\lambda-\lambda_1)(\lambda-\lambda_2)\cdots(\lambda-\lambda_n) = \lambda^n - (\lambda_1+\lambda_2+\cdots+\lambda_n)\lambda^{n-1} + \cdots + (-1)^n\lambda_1\lambda_2\cdots\lambda_n.$$

Exercise 11.5 The characteristic vectors belonging to λ are exactly the nonzero vectors in the null space of $\lambda I - A$. If $\lambda I - A$ has rank r, then by Theorem 10.6, its null space has dimension $n - r$. Hence we can find $n - r$ linearly independent nonzero vectors in this null space, and each belongs to λ. There cannot be more than $n - r$ such linearly independent vectors, otherwise the null space would have dimension greater than $n - r$.

Exercise 11.6 We have (using Exercise 5.7, generalized)

$$|\lambda I - C| = \begin{vmatrix} \lambda I - A & 0 \\ 0 & \lambda I - B \end{vmatrix} = |\lambda I - A||\lambda I - B|.$$

Thus the characteristic roots of C are $\lambda_1, \ldots, \lambda_m, \lambda_1', \ldots, \lambda_n'$, where $\lambda_1, \ldots, \lambda_m$ are those of A, and $\lambda_1', \ldots, \lambda_n'$ those of B. If $A\mathbf{v} = \lambda_1\mathbf{v}$, then

$$C\begin{bmatrix}\mathbf{v}\\ \mathbf{0}\end{bmatrix} = \lambda_1\begin{bmatrix}\mathbf{v}\\ \mathbf{0}\end{bmatrix},$$

and so on.

Exercise 11.9 By Theorem 7.13, the matrix A is singular if and only if there is a nonzero vector $\mathbf{v}$ such that $A\mathbf{v} = \mathbf{0}$. But this occurs if and only if 0 is a characteristic root of A, by Theorem 11.5.

Another approach: A is singular if and only if $\det A = 0$. Then use Exercise 11.4.

Exercise 11.10 As P ranges over the points $(x,0)$ on the X-axis, the point P' traces out the line with parametric equations

$$\begin{cases} x' = 2x \\ y' = 0, \end{cases}$$

that is, P' also traces out the X-axis.

As P ranges over all points $(0,y)$ on the Y-axis, we have for P':

$$\begin{cases} x' = 3y \\ y' = 3y, \end{cases}$$

so P' traces out the line $x' = y'$.

If $x + y = 7$, then

$$x' = 2x + 3y = 2(7 - y) + 3y = 14 + y, \qquad y' = 3y,$$

so $y = x' - 14$. Therefore

$$y' = 3(x' - 14)$$

is the equation of the line traced out by P' as P moves along the line $x + y = 7$.

We have

$$\overrightarrow{OP'} = \begin{bmatrix} x' \\ y' \end{bmatrix} = \begin{bmatrix} 2x + 3y \\ 3y \end{bmatrix} = \begin{bmatrix} 2 & 3 \\ 0 & 3 \end{bmatrix} \begin{bmatrix} x \\ y \end{bmatrix} = A \cdot \overrightarrow{OP},$$

where $A = \begin{bmatrix} 2 & 3 \\ 0 & 3 \end{bmatrix}$. If $A \cdot \overrightarrow{OP} = \lambda\, \overrightarrow{OP}$, then λ must be a characteristic root of A, so $\lambda = 2$ or $\lambda = 3$. For $\lambda = 2$, we find

$$A\mathbf{v}_1 = 2\mathbf{v}_1, \qquad \mathbf{v}_1 = \begin{bmatrix} 1 \\ 0 \end{bmatrix}.$$

Thus $\overrightarrow{OP'} = 2 \cdot \overrightarrow{OP}$ for P the point $(1,0)$. Consequently this also holds for *every* point P on the X-axis.

For $\lambda = 3$, we obtain

$$A\mathbf{v}_2 = 3\mathbf{v}_2, \qquad \mathbf{v}_2 = \begin{bmatrix} 3 \\ 1 \end{bmatrix}.$$

Thus $\overrightarrow{OP'} = 3 \cdot \overrightarrow{OP}$ for every point P on the line through $(0,0)$ and $(3,1)$.

Hence the mapping $P \to P'$ stretches all horizontal distances by a factor of 2, and all distances in the direction $\mathbf{v}_2$ by the factor 3.

Exercise 11.14 The equation $\mathbf{w}A = \lambda\mathbf{w}$ is equivalent to $A^T\mathbf{w}^T = \lambda\mathbf{w}^T$. Hence a nonzero solution $\mathbf{w}$ exists if and only if $\mathbf{w}^T$ is a characteristic vector of A^T belonging to λ. Therefore, λ must be a characteristic root of A^T, and hence also of A (by Exercise 11.13).

Exercise 11.16 Let $A\mathbf{v} = \lambda\mathbf{v}$. Then

$$A(B\mathbf{v}) = BA\mathbf{v} = B(\lambda\mathbf{v}) = \lambda(B\mathbf{v}).$$

Hence if $B\mathbf{v} \neq \mathbf{0}$, then $B\mathbf{v}$ is a characteristic vector of A belonging to λ.

Exercise 12.1

(a) $|\mathbf{a}| = 3$, $|\mathbf{b}| = 3$, $|\mathbf{a} + \mathbf{b}| = \sqrt{26}$. Angle between **a** and **b** is arccos 4/9.
(b) $\mathbf{c} = [x,y, \frac{1}{2}(x + 2y), \frac{1}{2}(2x + y)]$, x, y arbitrary.
(c) $\{\mathbf{u}_1, \mathbf{u}_2\}$ is an orthonormal basis, where

$$\mathbf{u}_1 = \frac{1}{3}[1,2,-2,0], \qquad \mathbf{u}_2 = \frac{1}{3\sqrt{65}}[14,1,8,-18].$$

(d) $\{\mathbf{w}_1, \mathbf{w}_2\}$ is an orthonormal basis, where

$$\mathbf{w}_1 = \frac{1}{3}[2,0,1,2], \qquad \mathbf{w}_2 = \frac{1}{3\sqrt{65}}[-8,18,14,1].$$

Exercise 12.2 See discussion in Theorem 13.7.

Exercise 12.4 Let

$$\mathbf{v}_1 = [0,1,0], \qquad \mathbf{v}_2 = [3,-1,4], \qquad \mathbf{v}_3 = [2,2,-1].$$

Then

$$\mathbf{u}_1 = \frac{1}{|\mathbf{v}_1|}\mathbf{v}_1 = \mathbf{v}_1, \qquad \mathbf{u}_1 \cdot \mathbf{v}_2 = -1, \qquad \mathbf{u}_1 \cdot \mathbf{v}_3 = 2.$$

$$\mathbf{v}_2' = \mathbf{v}_2 - (\mathbf{u}_1 \cdot \mathbf{v}_2)\mathbf{u}_1 = [3,0,4], \qquad \mathbf{u}_2 = \frac{1}{|\mathbf{v}_2'|}\mathbf{v}_2' = \frac{1}{5}[3,0,4].$$

$$\mathbf{v}_3' = \mathbf{v}_3 - (\mathbf{u}_1 \cdot \mathbf{v}_3)\mathbf{u}_1 - (\mathbf{u}_2 \cdot \mathbf{v}_3)\mathbf{u}_2 = \frac{11}{25}[4,0,-3], \qquad \mathbf{u}_3 = \frac{1}{5}[4,0,-3].$$

Exercise 12.6 Since $\{\mathbf{u}_1, \ldots, \mathbf{u}_k\}$ is a basis, we can express each **v** in V as a linear combination

$$\mathbf{v} = \alpha_1\mathbf{u}_1 + \cdots + \alpha_k\mathbf{u}_k.$$

Then

$$\begin{aligned}\mathbf{u}_i \cdot \mathbf{v} &= \mathbf{u}_i \cdot (\alpha_1\mathbf{u}_1 + \cdots + \alpha_k\mathbf{u}_k) = \alpha_1(\mathbf{u}_i \cdot \mathbf{u}_1) + \cdots + \alpha_k(\mathbf{u}_i \cdot \mathbf{u}_k)\\ &= \alpha_i.\end{aligned}$$

Exercise 12.8 Each vector **v** in V is of the form $\mathbf{v} = \alpha_1\mathbf{v}_1 + \cdots + \alpha_k\mathbf{v}_k$, where $\alpha_1, \ldots, \alpha_k$ are scalars. Then

$$\mathbf{w} \cdot \mathbf{v} = \mathbf{w} \cdot (\alpha_1\mathbf{v}_1 + \cdots + \alpha_k\mathbf{v}_k) = \alpha_1(\mathbf{w} \cdot \mathbf{v}_1) + \cdots + \alpha_k(\mathbf{w} \cdot \mathbf{v}_k) = 0.$$

Exercise 12.9 See proof of Theorem 13.7.

Exercise 13.1 Diagonal entries are $+1$'s and -1's.

Exercise 13.2

(a) $\begin{bmatrix} \cos 30° & -\sin 30° \\ \sin 30° & \cos 30° \end{bmatrix} = \begin{bmatrix} \frac{\sqrt{3}}{2} & \frac{-1}{2} \\ \frac{1}{2} & \frac{\sqrt{3}}{2} \end{bmatrix}$. (b) $\begin{bmatrix} -1 & 0 \\ 0 & 1 \end{bmatrix}$.

(c) $\begin{bmatrix} \cos\theta & \sin\theta \\ -\sin\theta & \cos\theta \end{bmatrix}$.

Exercise 13.6 Let $A = \begin{bmatrix} 1 & x \\ 0 & y \end{bmatrix}$. Then

$$I = AA^{\mathrm{T}} = \begin{bmatrix} 1 & x \\ 0 & y \end{bmatrix} \begin{bmatrix} 1 & 0 \\ x & y \end{bmatrix} = \begin{bmatrix} 1+x^2 & xy \\ xy & y^2 \end{bmatrix},$$

so $y = \pm 1$, $x = 0$.

Exercise 13.8 $$A^{-1} = \begin{bmatrix} \cos\theta & \sin\theta \\ -\sin\theta & \cos\theta \end{bmatrix}.$$

The transformation $\mathbf{v} \to A^{-1}\mathbf{v}$ is given by a clockwise rotation through angle θ about the origin.

Exercise 13.9 The transformation $\mathbf{v} \to AB\mathbf{v}$ can be accomplished in two steps: first $\mathbf{v} \to B\mathbf{v}$ (rotation through β), followed by $B\mathbf{v} \to A(B\mathbf{v})$ (rotation through α). Net result: rotate through $\alpha + \beta$.

Exercise 13.13 If $\mathbf{v}' = A\mathbf{v}$, then $\mathbf{v} = A^{-1}\mathbf{v}' = A^{\mathrm{T}}\mathbf{v}'$.

Exercise 13.14 By (13.6), we have

$$\begin{aligned} (A\mathbf{u}_i) \cdot (A\mathbf{u}_j) &= (A\mathbf{u}_i)^{\mathrm{T}}(A\mathbf{u}_j) = (\mathbf{u}_i^{\mathrm{T}}A^{\mathrm{T}})(A\mathbf{u}_j) \\ &= \mathbf{u}_i^{\mathrm{T}}(A^{\mathrm{T}}A)\mathbf{u}_j = \mathbf{u}_i^{\mathrm{T}}\,\mathbf{u}_j = \mathbf{u}_i \cdot \mathbf{u}_j. \end{aligned}$$

When $j = i$, this shows that $A\mathbf{u}_i$ is a unit vector. When $j \neq i$, this shows that $A\mathbf{u}_i$ is orthogonal to $A\mathbf{u}_j$.

Exercise 14.1

(a) $(x-3)^2 + 4(y+2)^2 = 16$.

Ellipse, center at $(3,-2)$, horizontal major axis of length 8, vertical minor axis of length 4.

(c) $y + \frac{5}{2} = -2x^2$.

Parabola, vertex at $(0,-\frac{5}{2})$, vertical axis of symmetry.

Exercise 14.2

(b) Let $\mathbf{u}_1 = \frac{1}{\sqrt{2}}\begin{bmatrix} 1 \\ -1 \end{bmatrix}$, $\mathbf{u}_2 = \frac{1}{\sqrt{2}}\begin{bmatrix} 1 \\ 1 \end{bmatrix}$, X'-axis along $\mathbf{u}_1$, Y'-axis along $\mathbf{u}_2$. Equation of conic in $X'Y'$-coordinate system is

$$-x'^2 + 3y'^2 = k.$$

This is a hyperbola if $k \neq 0$, and a pair of intersecting lines if $k = 0$.

Exercise 14.3

(b) Let $S = \begin{bmatrix} 0 & 1 & 1 \\ 1 & 0 & 1 \\ 1 & 1 & 0 \end{bmatrix}$. Characteristic roots are $-1, -1, 2$. Let

$$\mathbf{u}_1 = \frac{1}{\sqrt{2}}\begin{bmatrix} 1 \\ -1 \\ 0 \end{bmatrix}, \qquad \mathbf{u}_2 = \frac{1}{\sqrt{6}}\begin{bmatrix} 1 \\ 1 \\ -2 \end{bmatrix}, \qquad \mathbf{u}_3 = \frac{1}{\sqrt{3}}\begin{bmatrix} 1 \\ 1 \\ 1 \end{bmatrix}.$$

Then $S\mathbf{u}_1 = -\mathbf{u}_1$, $S\mathbf{u}_2 = -\mathbf{u}_2$, $S\mathbf{u}_3 = 2\mathbf{u}_3$. Let OX', OY', OZ' go in directions $\mathbf{u}_1$, $\mathbf{u}_2$, $\mathbf{u}_3$, respectively. Equation of quadric in $X'Y'Z'$-system is

$$-x'^2 - y'^2 + 2z'^2 = 8.$$

This is a hyperboloid of revolution about OZ'.

Exercise 15.1

$$\frac{\partial(F,G)}{\partial(x,y)} = \begin{vmatrix} F_x & F_y \\ G_x & G_y \end{vmatrix} = \begin{vmatrix} 2x & -2y \\ 2y & 2x \end{vmatrix} = 4(x^2 + y^2).$$

Exercise 15.2 In the diagram showing ρ, ϕ, θ, note that $OQ = \rho \sin \phi$, and $QP = \rho \cos \phi$. Then use

$$x = OQ \cdot \cos \theta, \qquad y = OQ \cdot \sin \theta, \qquad z = QP.$$

Exercise 15.3

$$\frac{\partial(r,\theta)}{\partial(x,y)} = \frac{1}{r}.$$

Exercise 15.5 We set

$$F(x,y,u,v) = xy - u - v, \qquad G(x,y,u,v) = x^2 - y^2 - 3u + v.$$

Then

$$\left(\frac{\partial y}{\partial u}\right)_{v=c} = -\frac{\dfrac{\partial(F,G)}{\partial(x,u)}}{\dfrac{\partial(F,G)}{\partial(x,y)}}.$$

$$\frac{\partial(F,G)}{\partial(x,u)} = \begin{vmatrix} F_x & F_u \\ G_x & G_u \end{vmatrix} = \begin{vmatrix} y & -1 \\ 2x & -3 \end{vmatrix} = -3y + 2x,$$

$$\frac{\partial(F,G)}{\partial(x,y)} = \begin{vmatrix} F_x & F_y \\ G_x & G_y \end{vmatrix} = \begin{vmatrix} y & x \\ 2x & -2y \end{vmatrix} = -2y^2 - 2x^2.$$

Exercise 15.6 Let

$$F = x^2 + yz - t^2 - u^2, \qquad G = y^2 - xz - tu, \qquad H = y + z - t - u.$$

Then

$$\left(\frac{\partial x}{\partial u}\right)_{t=c} = -\frac{\dfrac{\partial(F,G,H)}{\partial(u,y,z)}}{\dfrac{\partial(F,G,H)}{\partial(x,y,z)}} = \cdots.$$

Exercise 15.8 Let $F(x,y,z) = (x^2 + y^2) \cdot e^{xy} - z$. Then

$$\left(\frac{\partial x}{\partial z}\right)_{y=c} = -\frac{\dfrac{\partial F}{\partial z}}{\dfrac{\partial F}{\partial x}},$$

where $\dfrac{\partial F}{\partial z} = -1$, and

$$\frac{\partial F}{\partial x} = (x^2 + y^2) \cdot e^{xy} \cdot y + 2x \cdot e^{xy}.$$

Exercise 15.12 The closed region S_0 consists of the parallelogram with vertices (0,0), (2,0), (3,3), (1,3), and all points inside it. Further, $\dfrac{\partial(u,v)}{\partial(x,y)} = 6$ and

$$\iint_{S_0} (u + 4v)\, du\, dv = \iint_{R_0} (2x + 13y) \cdot 6\, dx\, dy.$$

Exercise 15.13 We have $y = \frac{1}{3}v$, $x = \frac{1}{2}u - \frac{1}{6}v$, $\dfrac{\partial(x,y)}{\partial(u,v)} = \frac{1}{6}$ and

$$\iint_{R_0} (2x - y)\, dx\, dy = \iint_{S_0} (u - \tfrac{1}{3}v - \tfrac{1}{3}v) \cdot \tfrac{1}{6}\, du\, dv.$$

Index